Perspektiven der Mathematikdidaktik

Series Editors

In der Reihe werden Arbeiten zu aktuellen didaktischen Ansätzen zum Lehren und Lernen von Mathematik publiziert, die diese Felder empirisch untersuchen, qualitativ oder quantitativ orientiert. Die Publikationen sollen daher auch Antworten zu drängenden Fragen der Mathematikdidaktik und zu offenen Problemfeldern wie der Wirksamkeit der Lehrerausbildung oder der Implementierung von Innovationen im Mathematikunterricht anbieten. Damit leistet die Reihe einen Beitrag zur empirischen Fundierung der Mathematikdidaktik und zu sich daraus ergebenden Forschungsperspektiven.

Anton Bastian

Teacher Noticing of Pre-service and In-service Secondary Mathematics Teachers

Influences of Teaching Experience, Cognitive Demands, and Teaching Internships on Perception, Interpretation, and Decision-making

Springer Spektrum

Anton Bastian
Fakultät für Erziehungswissenschaft
Universität Hamburg
Hamburg, Germany

Dissertation in fulfillment of the requirements for the academic degree of Doctor of Philosophy (Dr. phil.) at the Faculty of Education, Department of Social Sciences, Mathematics and Natural Sciences Education of the Universität Hamburg (University of Hamburg), submitted by Anton Bastian. Hamburg 2023.

ISSN 2522-0799 ISSN 2522-0802 (electronic)
Perspektiven der Mathematikdidaktik
ISBN 978-3-658-45618-4 ISBN 978-3-658-45619-1 (eBook)
https://doi.org/10.1007/978-3-658-45619-1

This Springer Spektrum imprint is published by the registered company Springer Fachmedien Wiesbaden GmbH, part of Springer Nature.
The registered company address is: Abraham-Lincoln-Str. 46, 65189 Wiesbaden, Germany

Preface

The present doctoral thesis by Anton Bastian, entitled Teacher Noticing of Pre-service and In-service Secondary Mathematics Teachers—Influences of Teaching Experience, Cognitive Demands, and Teaching Internships on Perception, Interpretation, and Decision-making, deals with current issues in research on the teaching profession and teacher education, which he conducted as part of follow-up studies to the international comparative study Teacher Education and Development Study in Mathematics (TEDS-M) within the TEDS research program. The thesis is based on work in the area of teachers' professional competence, and particularly teacher noticing, as it is embedded in the TEDS research program to use situation-specific constructs and video-based test instruments to get closer to classroom practice than it is possible with paper-and-pencil tests of teachers' professional knowledge. In the TEDS research program and in the present thesis, teacher noticing skills of pre-service and in-service teachers are conceptualized as a construct consisting of three facets, namely perception, interpretation and decision-making. Based on the teaching as a profession approach and the competence paradigm, this construct is understood as a mediator between dispositions, such as professional knowledge and beliefs, and performance, i.e., teaching and acting in classroom situations. This conceptualization is in line with the model of competence as a continuum formulated by Blömeke, Gustafsson and Shavelson in 2015. In the last two decades, the construct of teacher noticing has received increasing attention in the research community as it constitutes an essential component of teachers' professional competence and is a prerequisite for successful student-centered and knowledge-based teaching.

Anton Bastian's thesis builds on this relevant discourse and examines the differentiation of the construct of teacher noticing both theoretically and empirically. Furthermore, the thesis considers the current discourse on linking competence and expertise research by comparing novices and experts. Summarized, this work considers the following main questions:

- What is the structure, in other words the nature, of the construct teacher noticing?
- How does teacher noticing change, especially when experts are compared with novices to gain insight into possible developments, and how can possible changes be described in terms of professionalization?
- What key factors, such as time, professional experience, or learning opportunities, influence the development of teacher noticing?

In the light of these questions, the thesis is not only located in mathematics pedagogy, but also ties in with recent highly relevant questions about teacher noticing from general pedagogy and educational sciences as well as with methods and questions of quantitative empirical educational research. With this comprehensive approach, the thesis, which is based on three publications that address different research questions following the overall research interest and have been published in high-ranking peer-reviewed journals, achieves findings that go far beyond this individual work and points to further perspectives on the theoretical and empirical conceptualization and measurement of the construct of teacher noticing that may be investigated in the future.

The thesis is introduced with a general framework in which Anton Bastian takes up the current discourse on competence and expertise research among pre-service and in-service teachers. He develops and elaborates an extended impact model of teacher education and competence that combines approaches from supply-use models with impact models from research on the teaching profession and teacher education. The publications included in the thesis are then integrated into this model. Furthermore, the current state of the discussion on professional competence and teacher noticing is described and, due to the abundance of empirical studies, selected results based on eight relevant review studies are presented and discussed.

The methodological approach and the quality of the video-based instrument developed within the TEDS-M research program and used in this thesis are described. In order to examine the quality of the instrument, Anton Bastian analyzes correlations between the teacher noticing facets and the three commonly differentiated knowledge facets, and is thus able to demonstrate the strongly knowledge-based nature of the construct of teacher noticing.

The three publications included in the thesis, which focus on clearly distinguishable research questions, are then presented:

The first publication examines two questions, namely whether the theoretically distinct facets of teacher noticing can be empirically separated and measured with reliable scales, and whether teacher noticing differs between groups of pre-service and in-service teachers with different levels of teaching experience. Unexpected results emerge that can be interpreted as forgetting effects in teachers with long teaching experience.

The second publication builds on these group differences and examines the question of differences in teacher noticing between different groups of pre-service and in-service teachers with different levels of experience in relation to the cognitive demands of the practice-related items contained in the measurement instrument. Again, there emerge clear performance advantages for early-career teachers, particularly in the areas of decision-making, mathematics pedagogical knowledge, and cognitive demands in relation to recent development in mathematics pedagogy and general pedagogy.

The third publication addresses the question of how teacher noticing of pre-service teachers in their master's studies changes longitudinally as a result of extended practice phases (teaching internships) and how process variables (e.g., mentor support) influence the development of teacher noticing. The results demonstrate that teaching internships promote teacher noticing, with the facet of interpretation proving to be central, once again emphasizing the strong knowledge-based character of the construct.

Overall, Anton Bastian's thesis is highly relevant, informative, and methodologically advanced. The findings obtained advance the discourse on teacher noticing and its development both conceptually and methodologically. We hope that this work will be widely received and will inspire further studies.

Prof. Dr. Gabriele Kaiser
Prof. Dr. Johannes König

Acknowledgement

First of all, I would like to thank Prof. Dr. Gabriele Kaiser for her supervision and great support in writing this thesis and implementing the associated project, as well as for her guidance during my bachelor and master studies. She has always taken the time for my project and ideas and advised me with deep insight and with the big picture in mind, encouraging me to develop and progress.

I would also like to thank Prof. Dr. Johannes König for his methodological and strategic guidance in all of my publications, and him and Prof. Dr. Nils Buchholtz for reviewing my thesis. Jonas Weyers supported me in developing my methodological skills and challenged my conceptual understanding in discussions of the constructs of teacher noticing and professional vision, which I deeply appreciated and for which I thank him. I thank Natalie Ross for her content-related feedback and for critically discussing and developing new ideas with me. Armin Jentsch has always taken the time to help me with methodological problems and to answer my questions from all over the world, for which I am deeply grateful. I would like to thank Dennis Sommer for introducing me to the world of TEDS data and helping me with my first analyses.

Further thanks go to all the members of the Mathematics Education Research Group and its research colloquium at the University of Hamburg, especially Mireia Viguer Pérez, Ann-Kristin Glegoła, Karen Stadtlander, and Kirsten Benecke, and to Daniela Rehbock, Henri Weigand and Insa Kilian for their advice and (emotional) support. The great methodological advice, education, and feedback in the colloquium on quantitative methods in empirical educational research by Prof. Dr. Knut Schwippert also needs to be acknowledged. I would like to thank all my co-authors of my publications, particularly—next to those already mentioned persons—Prof. Dr. Björn Schwarz and Prof. Dr. Hans-Stefan Siller, for the fruitful collaboration and the constructive feedback.

A special thanks goes to my family, especially my partner, my parents and my brother, for their great, constant and unconditional support. Without them, I would not be where I am now.

Zusammenfassung

Vor dem Hintergrund von stetig wachsenden Anforderungen an Lehrkräfte und damit zusammenhängenden Bedarfen für die Untersuchung und Entwicklung unterrichtsbezogener Lehrkräftekompetenzen beschäftigt sich die vorliegende Dissertation mit situationsspezifischen Fähigkeiten, genauer der professionellen Unterrichtswahrnehmung, von Mathematiklehrkräften und Mathematiklehramtsstudierenden der Sekundarstufe. Mittels eines videobasierten Testinstruments wurden die professionelle Unterrichtswahrnehmung von Lehrkräften und Studierenden in den Facetten Wahrnehmung, Interpretation und Entscheidungsfindung aus pädagogischer und mathematikdidaktischer Perspektive gemessen, um so Erkenntnisse über Struktur und Charakteristika der Unterrichtswahrnehmung für verschiedene Lehrkräftegruppen sowie Einflüsse von Lehrerfahrung und Lerngelegenheiten zu erhalten. In zwei Teilstudien wurden 457 Teilnehmer*innen, bestehend aus Masterstudierenden, Junglehrkräften und erfahrenen Lehrkräften, querschnittlich in ihrer professionellen Unterrichtswahrnehmung getestet und verglichen, während sich eine dritte Studie mit 175 Masterstudierenden in einem längsschnittlichen Design genauer mit der Entwicklung der professionellen Unterrichtswahrnehmung und Einflussfaktoren auseinandersetzte.

Die Ergebnisse der Untersuchungen stützen die Konzeptualisierung des Konstrukts der professionellen Unterrichtswahrnehmung durch die drei Facetten Wahrnehmen, Interpretieren und Entscheidungentreffen und geben Aufschlüsse über die Beziehungen zwischen den drei Facetten. Im Vergleich der drei Erfahrungsgruppen zeigen sich weiter klare positive Einflüsse von Lehrerfahrungen für die Entwicklung der professionellen Unterrichtswahrnehmung. So zeigen Lehrkräfte deutlich höhere Fähigkeiten in allen Facetten als Masterstudierende, insbesondere in mathematikdidaktischen Themen. Allerdings erzielen erfahrenere Lehrkräfte ähnliche Ergebnisse wie Junglehrkräfte und weisen in Bereichen, die Mathematikdidaktik, das Entscheidungentreffen und neue Themen der mathematikdidaktischen und allgemeinpädagogischen Forschung betreffen, sogar signifikant schlechtere Leistungen als die Lehrkräfte mit weniger Unterrichtserfahrung auf. Dieses Ergebnis gibt Hinweise auf Sättigungs- oder Vergessenseffekte und verweist auf kulturelle Abhängigkeiten der professionellen Unterrichtswahrnehmung, da es im Kontrast zu vergleichbaren Studien aus ostasiatischen Kulturkreisen steht. Außerdem unterstreicht es die Notwendigkeit für adäquate Fortbildungsmöglichkeiten für erfahrene Lehrkräfte zur Entwicklung ihrer professionellen Unterrichtswahrnehmung. In der längsschnittlichen Betrachtung der professionellen Unterrichtswahrnehmung über die Zeit von Praxisphasen im Lehramtsmasterstudium hinweg konnte Interpretation als wichtiger Baustein in der Entwicklung von professioneller Unterrichtswahrnehmung ausgemacht werden, da diese Fähigkeiten auch die Entwicklung von Wahrnehmung und Entscheidungentreffen begünstigten. Damit wird der wissensbasierte Charakter von professioneller Unterrichtswahrnehmung betont. Des Weiteren zeigen Lerngelegenheiten wie die Verbindungen von in der Universität gelernten Theorien mit konkreten Situation in der schulischen Praxisphase und emotionale mentorielle Unterstützung positive Auswirkungen auf Facetten der professionellen Unterrichtswahrnehmung.

Stichwörter: professionelle Unterrichtswahrnehmung · Lehrkräftekompetenz · Lehrkräfteaus- und -fortbildung · (angehende) Mathematiklehrkräfte der Sekundarstufe · Lerngelegenheiten

Abstract

In light of steadily increasing demands on teachers and related needs for the investigation and development of teaching-related competences, this thesis examines the situation-specific skill of teacher noticing in pre-service and in-service secondary mathematics teachers. A video-based test instrument was used to measure pre-service and in-service teachers' noticing skills in the facets of perception, interpretation, and decision-making from general pedagogical and mathematics pedagogical perspectives to gain insight into the structure and characteristics of teacher noticing for different groups of teachers, as well as the influences of teaching experience and opportunities to learn. Two studies investigated and compared the teacher noticing skills of 457 participants consisting of master's students, early career teachers, and experienced teachers in a cross-sectional design, while a third study with 175 master's students in a longitudinal design examined the development of teacher noticing and the factors that influence it.

The results of the studies support the conceptualization of the construct teacher noticing as consisting of the three facets of perception, interpretation, and decision-making, and provide insight into the relationships among the three facets. A comparison of the three experience groups also revealed clear positive influences of teaching experience on the development of teacher noticing. For example, in-service teachers outperformed master's students in all facets, especially in the area of mathematics pedagogy. However, more experienced teachers performed similarly to early career teachers in general and even significantly worse in areas related to mathematics pedagogy, decision-making, and recent topics in mathematics pedagogical and general pedagogical research. This finding provided evidence for saturation or forgetting effects and indicated cultural dependencies in teacher noticing, as it contrasted with similar studies from East Asian cultural backgrounds. It also highlighted the need for adequate professional development opportunities for experienced teachers to develop their teacher noticing skills. An additional longitudinal study over the course of a teaching internship in the master's program found that interpretation was an important component in the development of teacher noticing, as these skills also facilitated the development of perception and decision-making. This emphasized the knowledge-based character of teacher noticing. Furthermore, opportunities to learn, such as linking theories learned at the university with concrete situations in the teaching internship and emotional support from mentors, showed positive effects on facets of teacher noticing.

Keywords: teacher noticing · teacher competence · teacher education · pre-service and in-service secondary mathematics teachers · opportunities to learn pre

List of Papers Included in This Thesis

Bastian, A., Kaiser, G., Meyer, D., & König, J. (2024). The link between expertise, the cognitive demands of teacher noticing and, experience in teaching mathematics in secondary schools. International Journal of Science and Mathematics Education, 22, 257–282. https://doi.org/10.1007/s10763-023-10374-x

Bastian, A., Kaiser, G., Meyer, D., Schwarz, B., & König, J. (2022). Teacher noticing and its growth toward expertise: An expert–novice comparison with pre-service and in-service secondary mathematics teachers. *Educational Studies in Mathematics, 110*(2), 205–232. https://doi.org/10.1007/s10649-021-10128-y

Bastian, A., König, J., Weyers, J., Siller, H.-S., & Kaiser, G. (2024). Effects of teaching internships on preservice teachers' noticing in secondary mathematics education. *Frontiers in Education, 9,* 1–15. https://doi.org/10.3389/feduc.2024.1360315

Contents

List of Figures

List of Tables

1 Introduction

Quality education has been identified as a key goal for society and sustainable development globally (United Nations [UN], 2015). As education comprises a complex system with numerous stakeholders, interests, and relationships, education systems are constantly facing new challenges, such as dealing with heterogeneity, transforming learning and teaching to represent a digitized world, and mitigating the effects of a global pandemic like COVID-19 (Helmke & Schrader, 2022; Köller et al., 2019; United Nations Educational, Scientific, and Cultural Organization [UNESCO], 2017). This requires not only the adaptability of the educational system as a whole but also the particular competence and diverse skills of teachers as key actors in student learning success (Hattie, 2009). The extensive requirements of the educational context challenge teachers to consider and incorporate in their teaching the diverse dispositions, understandings, and motivations of students, as well as the representations of the subject matter and the core components necessary for its comprehension (European Agency for Development in Special Needs Education, 2012; Köller et al., 2019; van Es & Sherin, 2002).

Furthermore, to facilitate learning, provide a rich and meaningful learning environment, and deliver high-quality instruction, teachers are required to manage multiple demands in complex situations of sensory overload at the moment of teaching (Klieme et al., 2001; Schlesinger et al., 2018; B. Sherin & Star, 2011). In particular, classroom management, individual learning support, and cognitive activation are necessary from a general pedagogical perspective, as well as subject- and teaching-related quality from a subject-specific perspective (Doyle, 1986; Klieme et al., 2001; Kounin, 1970; Schlesinger et al., 2018). Given the extensive

A. Bastian, *Teacher Noticing of Pre-service and In-service Secondary Mathematics Teachers*, Perspektiven der Mathematikdidaktik,
https://doi.org/10.1007/978-3-658-45619-1_1

demands on their teaching, teachers need to selectively attend to, understand, and act on relevant situations, which requires certain situation-specific skills that enable them to provide adaptive, high-quality instruction (Dindyal et al., 2021; Lachner et al., 2016; van Es & Sherin, 2002).

These situation-specific skills and their development have come into focus in educational research over the last two decades (Kaiser et al., 2017; van Es & Sherin, 2002). The construct of teacher noticing, often described as perceiving or attending to noteworthy events in the classroom, interpreting these events, or making sense of them, and sometimes also making subsequent decisions, has become a central topic within this research, as it promotes adaptive, student-centered teaching (Dindyal et al., 2021; König et al., 2022; Schack et al., 2017; M. G. Sherin, Jacobs, & Philipp, 2011a; van Es & Sherin, 2002). Since teacher noticing skills facilitate student achievement, enable teachers to act successfully in the classroom, and create learning opportunities, they constitute a crucial part of teachers' expertise and professional competence (Kaiser et al., 2017; Lachner et al., 2016; Metsäpelto et al., 2021; M. G. Sherin, Russ, & Colestock, 2011). Thus, educational research needs to focus on gaining insight into the nature of teacher noticing, its development, and its promotion in pre-service and in-service teachers. In addition, subject-specific findings are necessary, as teacher noticing includes both general pedagogical and subject-specific facets (Blömeke et al., 2016; van Es & Sherin, 2002). To inform policy makers and stakeholders in teacher education and professional development, a sound empirical basis and extensive research on the structure and development of teacher noticing and the influence of opportunities to learn in teacher education are needed. However, empirical quantitative research on teacher noticing at different stages of teachers' professional careers using standardized means of testing and large-scale studies remains scarce (König et al., 2022; Weyers, König, Santagata, et al., 2023).

1.1 Research Aims of the Study

In this thesis, I aimed to analyze teacher noticing skills in different groups of pre-service and in-service teachers, focusing particularly on the facets of perception, interpretation, and decision-making, comparisons between the groups of teachers, and the influence of opportunities to learn in teaching practice and experience. In doing so, I gathered insights and evidence about the structure, levels, and development of teachers' noticing skills in the context of German teacher education and development. The research presented in this publication-based thesis is divided into three publications, each dealing with specific aspects of my research interest. In particular, I addressed the following topics in the three core publications of my doctoral thesis:

A. The facet structure of teacher noticing
B. Expert–novice comparison between groups of teachers with different lengths of teaching experience
C. Analyses of the cognitive demands of teacher noticing in classroom situations
D. Demand-related differences between groups of teachers with different lengths of teaching experience
E. Longitudinal development of teacher noticing skills
F. Reciprocal interaction of facets of teacher noticing over time
G. Influences of teaching internships and their learning opportunities

In Publication I, I contributed to a study that investigated the structure of teacher noticing and explore expert–novice differences in a cross-sectional comparison, using teaching experience as a separating characteristic and thus addressing research subjects A and B. To follow up on this, I distinguished cognitive demands in noticing teaching situations to deepen the previous comparisons in Publication II (research subjects C and D). In Publication III, I took part in a study that complemented the results of the previous two publications with a longitudinal analysis of the development of teacher noticing and the influences of opportunities to learn in the background of the teaching internship to examine research subjects E, F, and G.

This thesis is organized into five main parts. First, I describe the context of this study and the research program within which it was conducted (Sect. 1.2), before characterizing the theoretical framework and summarizing the state of the art, focusing on professional competence, teacher expertise, and teacher noticing (Chap. 2). I then outline the chosen methodological approach, specifically explaining the instrument used to measure teacher noticing and providing evidence for its quality (Chap. 3). In the next chapters, the three publications that represent the core contribution of this thesis are summarized and described (Chaps 4, 5, and 6). Finally, I discuss the results of my thesis as a whole and conclude with an outlook on prospects and further developments (Chap. 7).

1.2 Context of the Study: The TEDS Research Program

In this section, I describe the research program that formed the context in which my thesis was conducted and in which the data collection took place. In the 2008 "Teacher Education and Development Study in Mathematics" (TEDS-M), an international team of researchers contrasted and compared primary and secondary mathematics teacher education programs in 17 countries. The study involved an investigation of pre-service teachers' competence at the end of their initial teacher education; in Germany, this meant at the end of the induction phase ("Referendariat"). These results were combined with curricular analyses and surveys of teacher educators to evaluate teacher education programs and to examine facets of professional competence in the form of teachers' professional knowledge and beliefs as well as opportunities to learn (Blömeke, Hsieh, et al., 2014; Tatto et al., 2012). Drawing from Shulman (1986, 1987), the TEDS-M framework distinguished three facets of teachers' knowledge: mathematics content knowledge (MCK), mathematics pedagogical content knowledge (MPCK), and general pedagogical knowledge (GPK; Blömeke & Kaiser, 2014). In addition, the study covered opportunities to learn in university teacher education and beliefs about mathematics and its teaching (Blömeke & Kaiser, 2014). The results from the TEDS-M study demonstrated the separability of the different knowledge facets and indicated strong disparities between the different countries (Blömeke & Kaiser, 2014). The TEDS-M study served as a foundation for several follow-up studies by the German group and thereby initiated the TEDS research program (for an overview, see Table 1.1), within which this thesis and project were conducted. In the following paragraphs, I provide a brief description of the projects whose data have been used in the context of this thesis.

Four years after the TEDS-M study, in the TEDS-Follow-Up study (TEDS-FU), German participants who were at the beginning of their teaching careers and had previously been part of the TEDS-M study were invited to participate in a follow-up study to assess the development of their professional competence (Kaiser et al., 2017). To come close to the classroom performance of the teachers, a video-based instrument to measure teacher noticing was developed within the frame of the study, which aimed to capture situation-specific facets of teachers' competence and to enrich and extend the cognition-oriented TEDS-M competence framework (Kaiser et al., 2015). The analyses confirmed the separability of teacher noticing skills from more cognitive facets of competence and, among other things, indicated the impact of teaching experiences on competence development and the importance of MCK and MPCK for situation-specific facets (Blömeke, König, et al., 2014; Kaiser et al., 2015).

In further development of the research program, the TEDS-Instruct (TEDS-I) and TEDS-Validate (TEDS-V) studies, student learning gains were included to fully evaluate the impact chain of teachers' competence on student learning progress. As an additional influential variable, teaching quality was included to better explain the relationship between the dispositions of teachers and their performance in the classroom, and to investigate the mediation effect on student learning. To assess instructional quality subject-specifically, researchers from the TEDS research program went beyond teacher competence and developed a new instrument evaluating instructional quality from general pedagogical and mathematics pedagogical perspectives, thus considering the interactions between competence and actual classroom performance (Schlesinger et al., 2018). Data from in-service teachers in four German federal states were included in TEDS-I and TEDS-V. Based on these data combined with student achievement tests, König et al. (2021) and Blömeke et al. (2022) studied the links between teacher competence, instructional quality, and student learning gains. For instance, they demonstrated teacher noticing skills as mediating effects of teachers' knowledge on instructional quality and student achievement, and thus as a proximal competence facet to classroom situations and student learning to student achievements (Blömeke et al., 2022).

To investigate the transferability of the teacher noticing instrument from in-service teachers to pre-service teachers and its applicability for longitudinal measurement, the TEDS-Validate-Transfer (TEDS-V-T) study measured the teacher noticing skills of pre-service teachers from six universities in Germany as well as self-assessed opportunities to learn in university teacher education and during the teaching internship, using extensive and detailed questionnaires. Weyers, König, Rott, et al. (2023) demonstrated the impacts of opportunities to learn and average high school diploma grades on teacher noticing skills and validated the use of the TEDS-FU teacher noticing instrument for measuring competence facets of pre-service teachers. To examine pre-service teachers' development of teacher noticing during their teaching internships at the end of their university education, a pre-post test design was employed. These change measurements are analyzed in the context of this thesis.

Table 1.1 Projects of the TEDS Research Program

Project abbreviation	Project name	Period	Target group	Main constructs investigated	Key publications
TEDS-M	Teacher Education and Development Study in Mathematics	2006–2010	PST	MCK, MPCK, GPK, beliefs, OTL	Blömeke et al., 2010a, 2010b
TEDS-Telekom	Teacher Education and Development Study—Telekom: Effectiveness of Innovative Mathematics Teacher Education Projects in the Subject of Mathematics	2008–2012	PST	MCK, MPCK, GPK, beliefs,	Buchholtz & Kaiser, 2013
TEDS-LT	Teacher Education and Development Study—Learning to Teach	2008–2012	PST	CK, PCK, GPK	Blömeke et al., 2011; Blömeke et al., 2013
TEDS-FU	Teacher Education and Development Study in Mathematics Follow-Up	2011–2013	IST	PID, MCK, MPCK, GPK, beliefs, OTL	Blömeke, König, et al., 2014; Kaiser et al., 2017; Kaiser et al., 2015
TEDS-I	Teacher Education and Development Study—Instruct	2014–2016	IST	PID, MCK, MPCK, GPK, beliefs, OTL	Felske et al., 2020; Jentsch et al., 2021; Schlesinger et al., 2018
TEDS-V	Teacher Education and Development Study—Validate	2016–2019	IST	PID, MCK, MPCK, GPK, beliefs, OTL	Blömeke et al., 2022; Kaiser & König, 2019; König et al., 2021

(continued)

Table 1.1 (continued)

Project abbreviation	Project name	Period	Target group	Main constructs investigated	Key publications
TEDS-East-West	Teacher Education and Development Study—East-West	2016–2018	PST, IST	PID, MCK, MPCK, GPK, beliefs, OTL	Yang, Kaiser, et al., 2021; Yang, König, & Kaiser, 2021
TEDS-V-T	Teacher Education and Development Study—Validate-Transfer	2019–2021	PST	PID, OTL	Bastian et al., 2022; Weyers, König, Rott, et al., 2023
TEDS-IME	Teacher Education and Development Study—Inclusive Mathematics Education	2022– 2024	PST, IST	PID-IT, MPCK-IT, GPK-IT	In progress

Note. PST—pre-service teachers, IST—in-service teachers, MCK—mathematics content knowledge, MPCK—mathematics pedagogical content knowledge, GPK—general pedagogical knowledge, OTL—opportunities to learn, CK—content knowledge, PCK—pedagogical content knowledge, PID—perception, interpretation, decision-making, IT—inclusive teaching. The table is based on a compilation by Weyers et al. (2022), and has been adapted and updated.

Theoretical Framework and State of the Art

2

In this section, I present the theoretical framework that forms the foundation of this thesis and provide an overview of the current state of research to contextualize my work and identify research desiderates. In the first section, I briefly delineate basic considerations of teaching and learning, and offer an expanded impact model of teacher education and competence, within which I later situate the three central publications of this thesis. Subsequently, I discuss teachers' professional competence and expertise as the main perspectives on teachers and the research on teachers utilized in this thesis. The construct of teacher noticing is then detailed and placed within the general framework of teachers' competence. Finally, I present a brief overview of teacher education and professional development in Germany.

2.1 Expanded Impact Model of Teacher Education and Competence

Facilitating student learning is a challenging task, as it is embedded in a complex interactional structure and influenced by various factors within multi-causal relationships (Helmke & Schrader, 2022). In this structure, and according to the supply-use model, teaching and instruction can be considered a supply, while the actual learning outcomes depend on the use of this supply, which in turn depends on various factors such as individual students' dispositions and instructional quality (Helmke & Schrader, 2022). In shaping instruction and thus creating quality and quantity of learning opportunities, teachers, as key stakeholders in education

A. Bastian, *Teacher Noticing of Pre-service and In-service Secondary Mathematics Teachers*, Perspektiven der Mathematikdidaktik,
https://doi.org/10.1007/978-3-658-45619-1_2

systems, play a vital role in facilitating student learning (Hattie, 2009; Yang & Kaiser, 2022).

Empirical studies have demonstrated the substantial influence and variance explanations of teachers and their competence on students' learning and achievement (Blömeke et al., 2022; Hattie, 2009; Hill et al., 2007; Lipowsky, 2006). Therefore, teachers and their competence are important explanatory variables for students' learning outcomes (Kaiser & König, 2019). Investigating the factors that strengthen teacher impact and promote teacher education and professional development constitutes a central field in educational research (Kaiser & König, 2019; König, 2020). As teachers' competence is also a product of learning opportunities, especially from their initial teacher education and professional development, it is not only an explanatory variable of student learning but also a dependent variable of teacher education (Helmke & Schrader, 2022; Kaiser & König, 2019). Thus, the relationship between teacher education, teachers' professional competence, and student learning can be understood and modeled in a twofold supply-use model, where the first supply consists of teacher education and professional development for teachers, and the second are the teachers, their competence, and their instruction for the students (Helmke & Schrader, 2022; Kaiser & König, 2019; Lipowsky, 2010; Yang & Kaiser, 2022).

To illustrate this twofold supply-use model and how the three publications included in this thesis are situated in the general context of educational processes, I adapted established impact models of professional teacher competence and their follow-ups (Blömeke, Gustafsson, & Shavelson, 2015; Blömeke & Kaiser, 2017; Kaiser et al., 2017; Krauss et al., 2020; Metsäpelto et al., 2021) and combined them with models of student learning outcomes (Helmke & Schrader, 2022) and of the impact of teacher education and professional development (Kaiser & König, 2019; Lipowsky, 2010; Yang & Kaiser, 2022). Based on these considerations, the structures of teacher education, professional competence, and student learning, as well as their relationships, can be represented through the expanded impact model of teacher education and teacher competence presented in Fig. 2.1. The model comprises four areas, each demonstrating one of the major steps of the twofold supply-use model discussed above: (1) teacher education and professional development, (2) teacher competence, (3) student learning activities, and (4) student outcomes. These are complemented by a fifth area, (5) context, to emphasize the strong impact of different cultures, regions, classes, and families on teaching and learning (Blömeke & Kaiser, 2017; Helmke & Schrader, 2022). In each area, important variables for the relationships among all five areas are presented. Note that I do not make claims about the completeness of this model and the variables it contains.

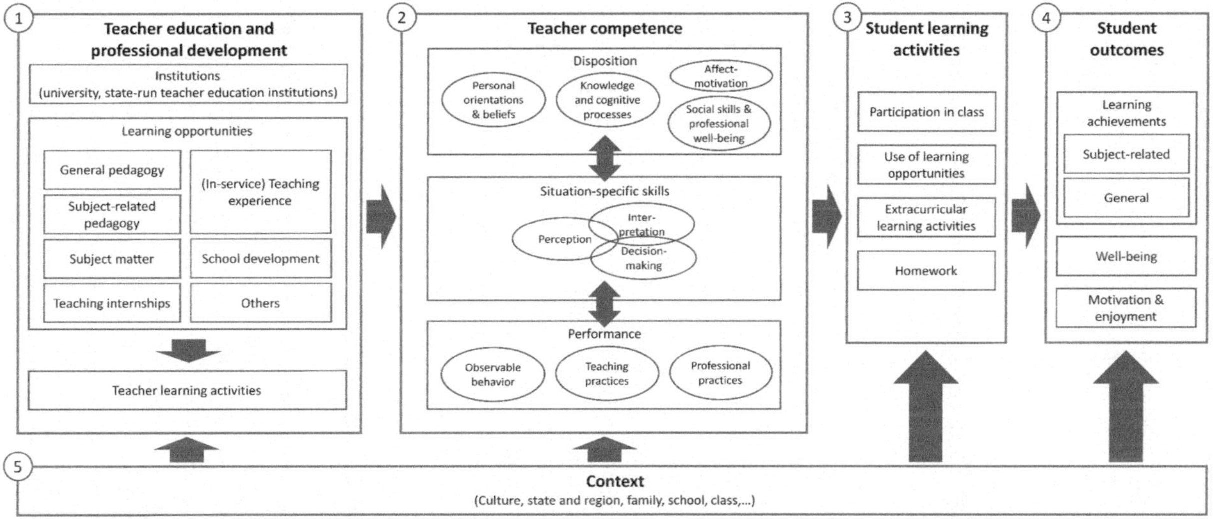

Note. Based on considerations by Blömeke, Gustafsson, and Shavelson (2015), Blömeke and Kaiser (2017), Kaiser and König (2019), Yang and Kaiser (2022), Krauss et al. (2020), Helmke and Schrader (2022), Metsäpelto et al. (2021), Kaiser et al. (2017), and Lipowsky (2010).

Fig. 2.1 Expanded Impact Model of Teacher Education and Competence

The following sections present the framework behind this expanded impact model and the central facets relevant to this thesis: professional competence and teacher expertise (Sect. 2.2), teacher noticing (Sect. 2.3), and teacher education (Sect. 2.4). As the thesis is situated in the context of professional competence and teacher education, the focus is on areas (1) and (2) of the expanded impact model.

2.2 Professional Teacher Competence and Expertise

In Germany, research on teachers, teaching, and teacher education is shaped by several perspectives and research traditions, including the personality approach, the structure–theoretical approach, the expertise approach, and the competence approach (Cramer et al., 2020; Terhart et al., 2014). The personality approach focuses on individual teachers' personalities and refers in particular to personality psychology and concepts of general personality traits, such as the Big Five model (Mayr, 2014). In contrast, structure–theoretical approaches consider the interactional structures of pedagogical performance and thus the structural logics of teaching and instruction, and reconstruct, for example, antinomies of teacher action (Helsper, 2007). Expertise and competence approaches in teacher education research, which are the most influential in the current discourse, are closely related; some might even say they are derivative of each other (König, 2020; Krauss, 2020). In the following section, I describe both approaches in more detail, starting with professional competence, and briefly contrast and compare them as they are applied in the three publications of this thesis.

2.2.1 Professional Competence

The construct of competence in the context of educational research and the professionalism of teachers has been in scholarly use for quite some time, but its conceptualization has only come to the forefront of educational and subject-specific pedagogical research in the last two decades (Koeppen et al., 2008; König, 2020). The study of teacher competence is closely related to the question of teaching as a profession and teachers' professionalization. The acceptance of teaching as a profession in the classical understanding of the term was established by Shulman's seminal work (1986, 1987, 1998), among others. With his definition of teachers as professionals, he also introduced a focus on teachers' professional knowledge as a crucial part of the teaching profession and distinguished between types of knowledge, such as content knowledge, general pedagogical knowledge, and pedagogical content knowledge (Shulman, 1986). This distinction strongly shaped subsequent competence research, particularly in mathematics education, and was implemented in the frameworks of several large-scale studies, for example, "Cognitive Activation in the Mathematics Classroom and Professional Competence of Teachers" (COACTIV) in the German context and TEDS-M in an international context (Blömeke, Hsieh, et al., 2014; Kunter et al., 2013).

In addition, the conceptualization of competence in these and subsequent studies was influenced by Weinert (2001), who defined competence as the available and learnable cognitive skills for successfully managing complex professional demands, as well as the motivational, volitional, and social willingness to apply these skills successfully and responsibly in various situations. Internationally, the definition of Koeppen et al. (2008), who described competence as "context-specific cognitive dispositions that are acquired and needed to successfully cope with certain situations or tasks in specific domains" (p. 62), emphasizing the situation-specificity of competence, is also in use (Kaiser & König, 2019). Both definitions are based on the assumption that competence is learnable, which in the case of teacher competence, describes teaching education as the acquisition of professional knowledge and skills and not primarily as nurturing talent (Baumert & Kunter, 2006; König, 2020).

To study teacher competence based on these definitions, various models have been developed to describe the construct (Baumert & Kunter, 2006; Blömeke, Gustafsson, & Shavelson, 2015). Studies in this context consider different aspects, called components (Baumert & Kunter, 2006; Blömeke, Gustafsson, & Shavelson, 2015) or facets (Kaiser et al., 2017), of teacher competence, such as areas of professional knowledge, professional orientations, motivational aspects, and situation-specific skills (Kaiser et al., 2017; König, 2020; Metsäpelto et al., 2021). The goal of these conceptualizations is usually not to directly operationalize the entire construct of teacher competence but rather to conduct operationalizations for selected facets of competence to test their separability and to empirically analyze them and their relations to other variables, and then to integrate the results into an overall model of competence (König, 2020). Therefore, the overall, usually generic models of competence often have heuristic and explanatory functions in particular (Baumert & Kunter, 2006; König, 2020).

Over the last two decades, research on teacher competence has primarily focused on teachers' dispositions, that is, cognitive abilities and affective-motivational characteristics, especially professional knowledge and beliefs

(Kaiser et al., 2017). This is reflected in the competence models of large-scale studies, such as TEDS-M and COACTIV (for details see Baumert & Kunter, 2013; Döhrmann et al., 2012). These models and the underlying conceptualizations specifically named knowledge, affects, and motivations as components of professional competence and separate them further, in particular incorporating the distinction of professional knowledge facets described by Shulman (1986).

This understanding of teacher competence was developed further by adding more context-specific facets that were already evident in Koeppen et al.'s (2008) definition. Since knowledge and other dispositions do not translate directly into classroom performance (Blömeke et al., 2022; Yang et al., 2020), situation-specific competence facets have been modeled that mediate dispositions and performance to complete the conceptualization of competence (Kaiser et al., 2017). With their model of competence as a continuum, Blömeke, Gustafsson, and Shavelson (2015) combined teachers' dispositions and successful action in classroom situations as constituents of competence, which had often been examined separately in prior research (Blömeke, Gustafsson, & Shavelson, 2015; Depaepe et al., 2013), by conceptualizing competence as a horizontal continuum between dispositions and performance and integrating both constituents of competence in their model. Between the two, they introduced situation-specific facets, namely perception, interpretation, and decision-making, as mediating and connecting dispositions such as knowledge with situations in the classroom (Blömeke, Gustafsson, & Shavelson, 2015). This model has been applied in various studies and adapted in further competence models (Krauss et al., 2020; Leuders et al., 2018; Metsäpelto et al., 2021; Santagata & Yeh, 2016). It can be understood as a foundation for extant research on the professional competence of teachers. Following this model, researchers have incorporated situation-specific measurement instruments into their studies, often using written or video vignettes as stimulus material (Dindyal et al., 2021; Santagata et al., 2021). In the TEDS research program, beginning with the TEDS-FU study (Kaiser et al., 2017; Kaiser et al., 2015), the model of competence as a continuum and situation-specific facets has been integrated, framing perception, interpretation, and decision-making as teacher noticing and thus relating and connecting competence research to the study of teacher noticing.

Empirical investigations have demonstrated the separability of different competence components and their facets, such as MCK, MPCK, and GPK as facets of professional knowledge (Blömeke & König, 2010b; König et al., 2018; Krauss et al., 2008), dispositional and situation-specific facets (Blömeke et al., 2016; Copur-Gencturk & Tolar, 2022; König et al., 2014), and beliefs (Voss et al.,

2011; Yang et al., 2020). Moreover, correlations and effects between the competence components have been explored, and several studies have shown positive correlations between the facets of professional knowledge, teacher noticing skills, and teachers' beliefs (Blömeke et al., 2016; Dreher & Kuntze, 2015; Meschede et al., 2017). Knowledge seems to positively influence situation-specific skills (Blömeke, König, et al., 2014; Jong et al., 2021; König et al , 2014), and these skills affect instructional quality and students' learning outcomes (Blömeke et al., 2022). Examining the other direction from teacher noticing back on teachers' dispositions, Santagata and Yeh (2016) illustrated the role of situation-specific facets in creating new knowledge and therefore influencing dispositions.

These results can be considered a justification for this way of modeling teacher competence. However, the structure of situation-specific skills, which are commonly framed as teacher noticing and which have been conceptualized through different facets and processes at the theoretical level (König et al., 2022; M. G. Sherin, Jacobs, & Philipp, 2011b), has received insufficient empirical attention in terms of discriminability (Seidel & Stürmer, 2014). Whether perceptual and interpretive facets can be empirically distinguished remains an open question, as they have often been described as strongly related and somewhat overlapping (Dreher & Kuntze, 2015; M. G. Sherin, Jacobs, & Philipp, 2011b; van Es & Sherin, 2008). Furthermore, interpretation and decision-making are sometimes pooled in empirical analyses (Yang, König, & Kaiser, 2021). This calls for an investigation of the separability of perception, interpretation, and decision-making as facets of teacher noticing.

2.2.2 Teacher Expertise

Following studies on expertise in general and in other domains, research on teacher expertise and expert teachers was introduced into the educational discourse in the 1980s by Berliner (1986, 1988) and others, and established in the German context by Bromme (1992). Since then, the expertise approach has been applied in various studies and is seen as highly influential in the educational sciences (Anderson & Taner, 2023; Krauss, 2020; Stigler & Miller, 2018). One of the main interests of research has been the identification of expert teachers and their characteristics, meaning not character traits but aspects of their practice, classroom performance, or dispositions (Anderson & Taner, 2023; Ericsson, 2018; Krauss, 2020). In general, defining experts has proven difficult, not only in the domain of teaching, and different approaches have been applied. Ericsson (2018) highlighted the importance of skills and knowledge that distinguish experts from novices in a domain, and identified the reproducibility of superior performance as a decisive characteristic in representative tasks for the domain. Hence, being an expert not only involves possessing knowledge and skills but also consistent and continuous skilled performance in the professional tasks of the respective domain (Winegard et al., 2018). This illustrates a performance-oriented perspective on expertise that also takes cognitions into account. Other perspectives have emphasized the social construction of expertise and the socialization processes in the development of expertise, with expertise defined as "the outcome of successful socialization" (Collins & Evans, 2018, p. 24) and "inseparable from individual and collective activities, concerns, equipment, and endeavors in particular domains" (Dall'Alba, 2018, p. 34). Thus, individual characteristics and social norms seem to be involved in expertise and the identification of experts. In addition, the possession of expertise has been shown to be domain specific; that is, experts have expertise in limited, well-defined domains, such as certain professions (Berliner, 2001; Ericsson, 2018).

In the domain of teaching, defining expertise and identifying expert teachers is challenging, because teaching and learning are embedded in a complex system of interacting elements, and exemplary performance is not easily identified or agreed on (Palmer et al., 2005; Stigler & Miller, 2018). Therefore, expertise definitions are commonly based on distinctions from people with less expertise, such as novices (i.e., beginners in a domain), or people from other domains (Berliner, 1988). Further, they are often based on prototypical descriptions that outline what features an expert teacher possesses and include characteristics that are often shared by experts in empirical studies (Anderson & Taner, 2023; Berliner, 2001; Sternberg & Horvath, 1995). However, to study expertise and arrive at empirical descriptions, experts must first be identified, which carries the risk of circular reasoning. For the identification of experts, teaching experience is often used as a decisive factor, as it has been shown to be an important condition for the development of expertise (Berliner, 2004; Bromme, 1992). However, experience alone seems to be insufficient to elicit expertise (Stigler & Miller, 2018), so additional criteria have been proposed and applied, such as social recognition, belonging to a professional group, or showing outstanding performance (Caspari-Sadeghi & König, 2018; Palmer et al., 2005). Student achievements are considered an important denominator of outstanding performance in the teaching context, but they have rarely been employed due to the associated survey effort and resource limitations (Caspari-Sadeghi & König, 2018; König et al., 2022; Stigler & Miller, 2018).

Another approach to describing expertise has been to study its development and define its particular levels. In his seminal work, Berliner (1986, 1988, 2004) adapted the general five stages of skill acquisition toward expertise by Dreyfus and Dreyfus (1986) for the teaching profession and described five stages of expertise development: novice teachers, advanced beginners, competent teachers, proficient teachers, and expert teachers. Novices base their teaching heavily on general rules, draw on situation-unspecific facts and knowledge, and apply them inflexibly in the classroom. Advanced beginners, who already have some initial teaching experience, begin to build strategic knowledge for teaching but have difficulty perceiving teaching situations selectively. Competent teachers, who have more teaching experience and thus case knowledge, possess these skills, make their decisions deliberately, and perform more situation-specifically in the classroom but often lack flexibility in the face of unexpected events. Proficient teachers can act more intuitively based on a rich body of case knowledge and rapid pattern recognition. Expert teachers can be characterized by their holistic, subconscious capture of classroom events and a flexible, in-the-moment selection of suitable instruction methods and reactions to students' actions and thinking. Berliner (1988, 2004) suggested that teachers need at least five to ten years of teaching experience to achieve expertise in teaching but also emphasized that few teachers will reach this level.

Empirical characteristics of expert teachers can be drawn from an extensive body of research, particularly for secondary mathematics teachers (Anderson & Taner, 2023). Early research demonstrated that expert teachers are exemplary in their ability to monitor and make sense of classroom interactions, as they possess a rich body of cognitive schemata for interpreting classroom events in meaningful ways (Carter et al., 1988; Sabers et al., 1991). In a metasummary (i.e., a systematic literature review, synthesis, and summary of studies with diverse methodology) of 106 articles from 1983 to 2021, Anderson and Taner (2023) identified 182 different characteristics that had been empirically linked to expert teachers in six domains: knowledge base, cognitive processes, beliefs, personal attributes, professionalism, and pedagogic practice. For example, the metasummary identified extensive, highly integrated knowledge in all domains, especially pedagogical content knowledge and knowledge about learners, cognitive schemata and means of automatization, constructivist beliefs, passion for teaching, extensive reflection, flexibility, and cognitively activating teaching as characteristics of expert teachers. In addition, expert teachers' instruction seems to be characterized by a balance between structure and improvisation, demonstrating the importance of building routines, and at the same time developing flexibility, adaptivity in teaching, and intuition (Anderson & Taner, 2023; Berliner, 1988).

Many of the attributes associated with expert teachers relate to noticing and performance in the moment of teaching (Anderson & Taner, 2023; Carter et al., 1988; Sabers et al., 1991). For instance, expert teachers are able to perceive large amounts of relevant information simultaneously, interpret it in detail, and make informed, appropriate decisions in the classroom (Anderson & Taner, 2023; Carter et al., 1988; Krull et al., 2007; Sabers et al., 1991). Research on expertise has often addressed processes that have also been examined in the research on situation-specific skills and teacher noticing; thus, it can be seen as a precursor within the discourse on teacher noticing skills (König et al., 2022; Krauss et al., 2020; Lachner et al., 2016). However, research on teacher noticing from an expertise-related perspective remains scarce, and in particular, the development of expertise in teacher noticing has only been examined in a few studies (König et al., 2022). Therefore, this research gap will be addressed in this thesis.

2.2.3 Comparative Reflections of Professional Competence and Expertise

The competence and expertise approaches are related but different perspectives on the research on teachers and teaching that both seek to describe, in common terms, "good" teaching, "good" teachers, and what characteristics or education teachers need to reach this level. As detailed above, they appear to be strongly related, and research following these approaches has often addressed similar aspects, such as professional knowledge, beliefs, in-the-moment skills or situation-specific skills, and classroom performance, although they may differ in their constructs and conceptualizations (Anderson & Taner, 2023; Blömeke, Hsieh, et al., 2014; Kunter et al., 2013). However, a distinct contrast can be noted between models of teachers' impact on students' achievement and the competence-as-continuum model by Blömeke, Gustafsson, and Shavelson (2015). Research from the competence approach usually starts with teachers' dispositions, particularly cognitions (Blömeke, Hsieh, et al., 2014; Kunter et al., 2013), and more recently, teachers' situation-specific skills (Kaiser et al., 2017), and examines their impact on classroom performance or student outcomes. In contrast, research using the expertise approach typically investigates actual exemplary performance, pedagogical practices, and their characteristics before looking at what might have caused them in terms of teachers' skills (Anderson & Taner, 2023; Krull et al., 2007; Sabers et al., 1991).

In summary, research on teachers' competence tends to examine its object of study from the perspective of teachers' dispositions, while research on teachers' expertise seems to study it from the other direction, starting from classroom performance and pedagogical practices. Since situation-specific skills are usually located in the middle, between dispositions and performance (Blömeke, Gustafsson, & Shavelson, 2015), combining or at least considering both approaches in research on teacher noticing promises to provide a more comprehensive view and support the acquisition of new insight into the characteristics of teacher noticing.

2.3 Teacher Noticing

In a literal sense, noticing is something that everyone does when they come into contact with any environment (Mason, 2002). However, to perform successfully in the classroom, a teacher must perceive classroom activities, attend to those events that are important for teaching and facilitating learning, make sense of these events, and integrate their interpretation into further teaching (Dindyal et al., 2021; van Es & Sherin, 2002). These challenges require special, sensitized, and professional noticing skills, often referred to as teacher noticing (Mason, 2002; M. G. Sherin, Jacobs, & Philipp, 2011b). In recent years, teacher noticing has received increased attention in educational research, especially in mathematics education, as can be seen in many anthologies and special issues of journals (e.g., Dindyal et al., 2021; Schack et al., 2017; M. G. Sherin, Jacobs, & Philipp, 2011a) as well as in recent literature surveys (Amador et al., 2021; König et al., 2022; Wei et al., 2023). Researchers have begun to examine what characteristics this construct encompasses (M. G. Sherin, Jacobs, & Philipp, 2011b; van Es & Sherin, 2002), how it can be measured (Kaiser et al., 2015; Seidel & Stürmer, 2014), what noticing skills pre-service and in-service teachers possess (Jacobs et al., 2010), how the construct is connected to other variables (Blömeke, König, et al., 2014; König et al., 2014), how teacher noticing develops over time (Jong et al., 2021), and how teacher noticing skills can be promoted (M. G. Sherin, 2007). These research areas are informed by different theoretical perspectives on teacher noticing, which I describe in the next section.

2.3.1 Theoretical Perspectives on Teacher Noticing

Considering the extensive body of research on teacher noticing, its conceptualization in different studies and research projects remains heterogeneous, as different perspectives have shaped the field. In an extensive literature survey, König et al. (2022) distinguished four main influential approaches: cognitive-psychological, expertise-related, sociocultural, and domain specific. In addition, Scheiner (2021) proposed an embodied-ecological view of teacher noticing to complement these four perspectives, which can be understood as an emergent approach in the field (Kaiser et al., 2023).

The most influential perspective identified is the cognitive-psychological approach, particularly informed by the work of van Es and Sherin (2002), which focuses on analytically distinguishing and investigating facets of teacher noticing to reconstruct the (partially subconscious) cognitive processes that teachers engage in during teacher noticing (König et al., 2022). These processes often include perceiving or attending to particular noteworthy events in the classroom, making sense of or interpreting these events, and sometimes making subsequent decisions (König et al., 2022; Scheiner & Kaiser, 2023). Additionally, some studies include the enactment of these decisions in their considerations (Thomas et al., 2020). The aim of studies following this approach is often to promote or measure pre-service teachers' noticing skills in relation to the respective cognitive processes involved, or to gain insight into what and how they perceive, and the interpretation or handling of instructional events (Santagata et al., 2021). This approach includes two main frameworks in research on teacher noticing (Amador et al., 2021; König et al., 2022): the *Learning to Notice* framework (van Es, 2011) and the *Professional Noticing of Children's Mathematical Thinking* framework (Jacobs et al., 2010).

Another perspective, referred to as the expertise paradigm, which also investigates cognitive processes but focuses more on differences between novices and experts, can be seen as a precursor to research on teacher noticing (Lachner et al., 2016), establishing an expertise-related perspective on teacher noticing. For example, Berliner (1988), who distinguished stages of expertise development, and Carter et al. (1988), who investigated differences between experts' and novices' perceptions, informed this approach although they did not specifically use the term "teacher noticing." Recent studies following this approach include expert–novice comparisons of visual expertise and knowledge-based reasoning (Gegenfurtner, 2020) or expertise-dependent profiles of teacher noticing (Jacobs et al., 2022).

The sociocultural perspective is characterized by an emphasis on the situation-specificity and social construction of teacher noticing (König et al., 2022). This builds on the work of Goodwin (1994), who investigated the construct of professional vision,[1] conceptualizing it as "socially organized ways of seeing and understanding events that are answerable to the distinctive interests of a particular social group" (p. 606). This incorporates the relevance of culturally based values as well as shared norms of communities of practice, which are to be understood as socioculturally formed groups, for the development of individual noticing skills into the understanding of teacher noticing. The term "professional vision" has been widely adopted (Seidel & Stürmer, 2014; M. G. Sherin, 2007), but often with a cognitive-psychological basis and without adapting the sociocultural implications of this approach (Santagata et al., 2021; Scheiner & Kaiser, 2023).

Another perspective distinguished is the discipline-specific approach, which is grounded in the work by Mason (2002, 2011), who focused on fostering a sensitized and methodical awareness of aspects of one's own teaching practice, as well as of students and their ideas. This places the approach, which is more focused on individual teachers and their development, at a less directly practice-oriented level than other approaches. He defined teacher noticing in this sense as "a collection of practices designed to sensitize oneself so as to notice opportunities in the future in which to act freshly rather than automatically out of habit" (Mason, 2011, p. 35). This approach emphasizes the role of individual teachers and the development and training of their cognitive skills to apply certain practices to make teacher noticing intentional, and promotes a shift from retrospective to in-the-moment noticing (Mason, 2002; Scheiner & Kaiser, 2023).

A recent emergent approach by Scheiner (2021) complements these four perspectives. Drawing from Gibson's (1979) ecological approach to perception, the embodied-ecological approach emphasizes the active, exploratory role of percipients in the act of perceiving and considers how teachers engage with the classroom environment, such as through eye or body movements, to perceive things and gain information (Scheiner, 2021). Thus, the perceiver and the environment are in a reciprocal, interdependent relationship, emphasizing the active role of the perceiver in what they perceive (Scheiner & Kaiser, 2023). This emergent perspective is reflected in new research directions in the field (Scheiner & Kaiser, 2023). For example, Kosko et al. (2021) explored how pre-service teachers use head movements to interact with 360° videos of classroom situations,

[1] In the recent state of the art, the terms teacher noticing and professional vision are understood to describe similar and highly overlapping constructs (Martin et al., 2022; Weyers, König, Santagata, et al., 2023). In this thesis, I use the more common term teacher noticing to describe the construct in our own conceptualization (König et al., 2022), but apply the authors' terminology when referring to specific research and the related frameworks.

and van Es and Sherin (2021) introduced shaping (i.e., constructing interactions to collect additional information) as a new facet in their conceptualization of teacher noticing.

2.3.2 Conceptualization of Teacher Noticing in the TEDS Research Program

In the TEDS research program, the construct of teacher noticing has been incorporated since the TEDS-FU study connected to the development of the model of competence as a continuum and of a new video-based measurement instrument (Blömeke, Gustafsson, & Shavelson, 2015; Kaiser et al., 2017; Kaiser et al., 2015). Referring to Blömeke, Gustafsson, and Shavelson's (2015) understanding of situation-specific skills as consisting of perception, interpretation, and decision-making, we conceptualize teacher noticing as consisting of the following three facets:

(a) perceiving particular events in an instructional setting; (b) interpreting the perceived activities in the instructional setting; (c) decision-making, either as anticipating responses to students' activities or as proposing alternative instructional strategies. (Kaiser et al., 2015, p. 374)

This definition is primarily influenced by the cognitive-psychological perspective on teacher noticing and combines perceptual, interpretive, and decision-related facets of in-the-moment noticing processes. Drawing on expertise research (Berliner, 1988; Carter et al., 1988) and thus integrating an expertise-related perspective on teacher noticing, the first facet consists of perceptual processes with no or only minor interpretation (Bastian et al., 2022; Yang, König, & Kaiser, 2021). This facet is focused only on observing clearly discernible incidents and facts in the classroom. This conceptualization is different from other descriptions, in which perception, called (selective) attention, and interpretation are much more closely connected (Dreher & Kuntze, 2015; van Es & Sherin, 2008). The second facet comprises making sense of and interpreting perceived incidents. Thus, it links specific situations to abstract, general knowledge, such as MCK, MPCK, and GPK; case knowledge and teaching experience; and beliefs and motivations, particularly constituting the mediation of teacher noticing between dispositions and actual performance. The third facet explicitly includes a decision-oriented facet to emphasize the performance- and action-oriented nature of teacher noticing (Kaiser et al., 2015), since noticing is strongly related to taking action, as described by Erickson (2011).

In terms of content or events in the focus of situation-specific skills, our conceptualization of teacher noticing not only focuses on students' mathematical thinking or a specific mathematical content alone, but encompasses a variety of noteworthy events relevant to quality mathematics education (e.g., Blum, 2019; Jentsch et al., 2021), such as individual students' mathematical thinking and misconceptions, cognitive activation, designing learning processes, and classroom management from mathematics pedagogical and general pedagogical perspectives (Kaiser et al., 2015; Yang, König, & Kaiser, 2021). Hence, this conceptualization is designed to facilitate a comprehensive understanding and investigation of teacher noticing.

2.3.3 Research on Teacher Noticing and Selected Empirical Results

Given the extensive body of research on teacher noticing, this section focuses on literature reviews[2] and summarizes important findings that directly relate to the research objective of my thesis. To date, there have been several reviews of studies on teacher noticing, each with special research interests and foci (ordered by year of publication):

1) Stahnke et al. (2016) specifically addressed the facets of perception, interpretation, and decision-making (60 articles included);
2) Grub et al. (2020) focused on the professional vision of classroom management with the application of eye-tracking methods (12 articles included);
3) Amador et al. (2021) analyzed studies of prospective teachers' noticing in the context of two common frameworks (49 articles included);
4) Chan et al. (2021) reviewed studies on teacher noticing in science education (29 articles included);
5) Santagata et al. (2021) reviewed video-based programs (35 articles included);
6) König et al. (2022) conducted a systematic literature review of all studies on teacher noticing and professional vision (182 articles included);
7) Wei et al. (2023) conducted a bibliometric review of teacher noticing (139 articles included); and

[2] I have not included overview articles in special issues (e.g., Dindyal et al., 2021) or edited thematic anthologies on teacher noticing (e.g., Schack et al., 2017; M. G. Sherin, Jacobs, & Philipp, 2011a), which, nonetheless, provide comprehensive information and insights into the research.

8) Weyers, König, Santagata, et al. (2023) focused on standardized instruments measuring teacher noticing (37 articles included).

First, it should be noted that research on teacher noticing is a steadily growing field (König et al., 2022). Wei et al. (2023) described exponential growth in the number of studies since 2008. In the recent discourse, studies using cognitive-psychological and thus analytical approaches, studies on pre-service teachers, and qualitative studies are particularly numerous (Amador et al., 2021; König et al., 2022; Wei et al., 2023). However, the research field has been shaped not only by a variety of perspectives in frameworks, as described in Sect. 2.3.1, but also by a variety of methods and study designs, such as qualitative, quantitative, and mixed-methods studies, as well as varying numbers of measured time points and cross-sectional designs, which can make it difficult to contrast and compare findings (Amador et al., 2021; König et al., 2022). In addition, researchers from North America and Europe dominate the field, so more contributions from other parts of the world and cross-cultural studies are needed to investigate the cultural characteristics of teacher noticing (König et al., 2022; Wei et al., 2023). The use of videos, either videos of teachers' own teaching, videos of others' instruction, or scripted videos as stimulus material in eliciting and measuring teacher noticing, is also characteristic of the field (Santagata et al., 2021; Wei et al., 2023; Weyers, König, Santagata, et al., 2023). Major domain-specific foci in research on teacher noticing include subject-specific pedagogical topics, such as student thinking or subject matter, and general pedagogical topics, such as instructional practices or classroom management (König et al., 2022; Weyers, König, Santagata, et al., 2023). To date, studies have focused primarily on mathematics or science education (Chan et al., 2021; König et al., 2022; Wei et al., 2023).

In terms of research findings, only reviews 1), 2), 3), 4), and 6) reported empirical results on teacher noticing. Stahnke et al. (2016) analyzed the reported influences on teacher noticing skills and concluded that there were effects of teachers' knowledge, beliefs, and goals on teacher noticing. In particular, knowledge and belief facets such as MPCK, MCK, and constructivist beliefs significantly predicted teacher noticing (Stahnke et al., 2016), and teacher noticing seemed to predict instructional quality (Stahnke et al., 2016). Expertise and teaching experience have been shown to facilitate noticing (König et al., 2022; Stahnke et al., 2016). In the few existing expert–novice comparisons, expert teachers demonstrated significantly better performance than novices on the noticing facets studied, with varying effect sizes (König et al., 2022; Meschede et al., 2017; Yang, König, & Kaiser, 2021). Novice teachers seem to have particular difficulties with decision-making and the application of teacher noticing in lesson phases that were not whole class discussions (König et al., 2022; Stahnke et al., 2016). Eye-tracking studies have shown that expert teachers are able to perceive more classroom events than novices and focus more on relevant aspects of the classroom situation (Grub et al., 2020). More research, especially quantitative studies, is needed to substantiate the findings on the effects of expertise and teaching experience on teacher noticing (König et al., 2022). This research gap is addressed in Publications I and II of this thesis (see Chaps. 4 and 5).

Another focus of research is the promotion of teacher noticing facets in teachers, especially pre-service teachers, via interventions (Amador et al., 2021; König et al., 2022). Studies have reported positive effects of interventions on attending and reasoning skills as well as on responding to student actions and proposing alternative instruction (König et al., 2022). In contrast, in an analysis of a subset of 27 studies, Amador et al. (2021) reported inconclusive results due to differences in research methods, frameworks, and study designs, but also highlighted the results of studies that demonstrated higher levels of noticing skills after implementing facilitative interventions. The small number of studies with more than one time point of measurement calls for more longitudinal study designs to examine the effects of interventions and the development of teacher noticing over time (Amador et al., 2021; König et al., 2022). This research gap is addressed in Publication III of this thesis (see Chap. 6).

2.4 Teacher Education and Professional Development

Since the three publications in this thesis study groups of teachers at different stages of their professional careers, and some teacher education programs in particular, I briefly describe teacher education in Germany in this section. German teacher education comprises three main phases (see Fig. 2.2): (1) higher education at universities or pedagogical universities, (2) practical teacher training (preparatory service or induction phase), which is carried out by special state-run teacher training institutions, and (3) elective professional development programs for in-service teachers (Drahmann, 2020; Eckhardt, 2021). The first two phases can be understood as initial teacher education, since participants are considered fully trained teachers upon their conclusion. The third phase is for in-service teacher education and professional development. Although the design and organization of teacher education depend on the specific federal state in which the education takes place (Cortina & Thames, 2013), they also share some common characteristics. For secondary education, which is the focus of this thesis, several types of schooling have been implemented: teaching in academic-track schools ("Gymnasium"), teaching in nonacademic-track schools, teaching in intermediate forms of these two, teaching in vocational schools, and special needs education (Cortina & Thames, 2013; Drahmann, 2020). With the exception of special needs teachers, all teachers are required to study at least two school subjects.

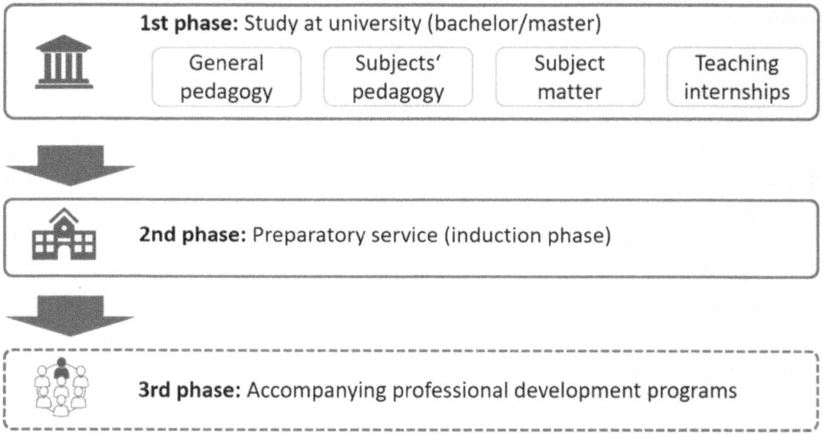

Fig. 2.2 Phases of teacher education in Germany

The first phase of teacher education is structured in a bachelor-master system and is divided into courses on general pedagogy, subject pedagogy, and subject matter (Terhart, 2021). This is supplemented by field experience, called teaching internships or practicums, to facilitate the transition from university to school, link theory to practice, and provide professional orientation (König, 2019). The first phase typically takes five to six years to complete.

In the preparatory service, teachers begin teaching in schools while being supervised by special state-run teacher training institutions, experienced teacher educators, and mentors from their school. The preparatory service consists of seminars, classroom observation, and accompanied and independent teaching, and lasts 1.5–2 years (Eckhardt, 2021).

The third phase of teacher education accompanies in-service teachers throughout their professional lives and consists of a variety of professional development programs, often developed and provided by the teacher training institutions of the respective federal state (Eckhardt, 2021). The programs are often led by experienced, specially trained teachers and cover a wide range of relevant topics, such as teaching, learning, and school development (Richter & Richter, 2020). Teachers can choose which programs they want to participate in but are required to attend some programs during the school year. Non-participation has no negative consequences, so participation rates are often low (Richter & Richter, 2020; Wermke, 2011).

Methodological Approach 3

In this section, I briefly describe the measurement of competence and, in particular, of teacher noticing from a general point of view before elaborating on the TEDS instrument for measuring teacher noticing as a facet of professional teacher competence. Then, the quality of the instrument and the validity of the test score interpretation for the three publications are examined.

3.1 Measurement of Competence and Teacher Noticing

3.1.1 Basic Principles of Competence Measurement

In general, competence can be understood as a latent construct often based on implicit theoretical assumptions that only become explicit through measurement or observation of action (Shavelson, 2010). Competence itself is not observable but only visible through the competent performance of individuals in concrete situations (Shavelson, 2010). Therefore, measuring competence in specific domains requires a precise definition of what competence entails to describe how it operates in specific situations (Klieme & Hartig, 2008). An example of a competence definition is the description by Weinert (2001) introduced in Sect. 2.2.1. Based on this definition, realistic tasks and measurement instruments can be developed to expose individuals to situations that can elicit competent behavior (Shavelson, 2010).

In his seminal work on competence testing, McClelland (1973, p. 7) stated that "the best testing is criterion sampling," meaning that tests should primarily

A. Bastian, *Teacher Noticing of Pre-service and In-service Secondary Mathematics Teachers*, Perspektiven der Mathematikdidaktik,
https://doi.org/10.1007/978-3-658-45619-1_3

focus not on cognitive structures but on the criteria of the situations in which the respective competence comes into action. Following this, Shavelson (2010) formulated some characteristics for quality items to measure competence: "tasks and responses should [...] (a) be real-life in nature, (b) tap complex abilities and skills, (c) be amenable to practice and improvement, and (d) be amenable to standard setting" (p. 46). However, such real-life testing of participants, which in the teaching context would mean observations of classroom instruction in particular, is often not feasible due to limited time and financial resources, especially when working with large samples and collecting quantitative data (Blömeke, Gustafsson, & Shavelson, 2015; Hughes & Huby, 2002; Piwowar et al., 2018). Moreover, to achieve high-quality psychometric measurements, competence models need well-defined facets that can be reliably operationalized while ensuring the validity of test score interpretation (Blömeke, Gustafsson, & Shavelson, 2015; Koeppen et al., 2008). To enable both a reliable assessment according to validity criteria and an economical assessment of competence in large-scale studies, competence measurement should combine knowledge and personality tests and situation-specific behavioral tests (Blömeke, Gustafsson, & Shavelson, 2015). For example, knowledge tests for MCK, MPCK, and GPK and a beliefs questionnaire were combined with situation-specific teacher noticing measures in the TEDS-V and TEDS-I studies (Blömeke et al., 2022; Kaiser et al., 2017).

As a consequence of this comprehensive testing with multiple individual scales, even measuring parts of the latent construct of competence often requires the use of complex statistical models on a test-theoretical level, such as models applying item response theory (IRT; Klieme & Hartig, 2008). This theory was specifically developed to use dichotomous items (i.e., items that can be correct or incorrect), which are often used in the measurement of competence to assess latent constructs (Kelava & Moosbrugger, 2011). This theory states that there is a positive relationship between an observed latent variable and the probability of solving an item testing the variable, marking IRT as a probabilistic theory (Bond et al., 2021; Kelava & Moosbrugger, 2011). One model that has been widely used in the context of IRT is the Rasch model (Rasch, 1960). It allows for the simultaneous estimation of person and item characteristics by attributing correct item responses to the person's ability in the considered latent construct and item difficulty (Bond et al., 2021; Klieme & Hartig, 2008). As a basic assumption, whether an individual answers an item correctly can be described by a probability function that depends only on the individual's ability and the difficulty of the item (Rasch, 1960). Furthermore, the Rasch model facilitates the construction of scales even when participants have not answered the exact same sets of items, enabling the use of balanced incomplete test designs such as those used in TEDS-M (Tatto, 2013).

3.1.2 Methods in the Study of Teacher Noticing

The study, empirical investigation, and measurement of teacher noticing in general (i.e., not just in the context of competence frameworks) have been heterogeneous to date (Dindyal et al., 2021; Santagata et al., 2021). Researchers have applied various methodological approaches following qualitative, quantitative, or mixed-method research designs: (1) using stimulus materials such as written or video-based artifacts of real or scripted classroom performance to elicit and measure teacher noticing through teachers' responses to specific prompts, (2) exposing teachers to their own or others' captured practice and asking them to share their noticing, (3) collecting retrospective remarks of teachers about their noticing in the classroom, and (4) observing teachers during classroom instruction (Jacobs & Spangler, 2017; König et al., 2022; M. G. Sherin, Russ, & Colestock, 2011; Weyers, König, Santagata, et al., 2023).

In quantitative, large-scale measurements of teacher noticing skills, researchers have primarily adopted the first approach and used selected artifacts of real or constructed school practice in their tests, which are referred to as vignettes (Weyers, König, Santagata, et al., 2023). When testing noticing skills, participants, who are usually pre-service or in-service teachers, are first presented with these vignettes as stimuli and then prompted to apply teacher noticing and create (often written) evaluable products of this process (Santagata et al., 2021; Weyers, König, Santagata, et al., 2023). Videos have played a crucial role in these vignettes, as demonstrated by recent literature surveys (Santagata et al., 2021; Weyers, König, Santagata, et al., 2023), because they are particularly suitable for simulating classroom situations, although written vignettes have also been used (Dindyal et al., 2021; Jacobs et al., 2010; Kramer et al., 2020; Piwowar et al., 2018; Santagata et al., 2021). Some instruments, such as the "Observer" (Seidel & Stürmer, 2014), or the "Professional Vision of Inclusive Classroom" instrument (Keppens et al., 2019), use recordings of real lessons that are selected from both theoretical and methodological perspectives and edited to fit the test. In contrast, scripted (i.e., staged) videos are included in the TEDS instrument for teacher noticing (Kaiser et al., 2015). The items used in the instrument are often open-ended, asking participants to write free or pre-structured text (Jacobs et al., 2010; Kaiser et al., 2015), and/or rating scale formats, requiring the participants to state their level of agreement with certain statements or to assess characteristics of the stimulus used (Gold & Holodynski, 2017; Kaiser et al., 2015; Seidel & Stürmer, 2014; Weyers, König, Santagata, et al., 2023). In some cases, special item formats are employed, such as comparative judgments between two video vignettes (Keppens et al., 2019).

3.2 TEDS Instrument for Measuring Teacher Noticing

3.2.1 Description of the Instrument

To date, while situation-specific skills have been integrated into conceptual frameworks, teacher noticing measurement instruments have seldom been integrated into measurement frameworks of comprehensive competence models at the empirical level, although there are some exceptions (Blömeke et al., 2022; Jong et al., 2021; Kaiser et al., 2017; Meschede et al., 2017). As part of the TEDS research program, a video-based instrument was developed and applied for the first time in TEDS-FU and then utilized in subsequent studies.[1]

The instrument comprises three scripted video vignettes depicting lessons in the ninth and tenth grades in academic-track and non-academic-track schools. The videos range from 2.25 to 3.5 min in length and cover a wide range of different teaching phases (e.g., lesson introduction, partner work, and whole class discussion of results) and mathematical topics (e.g., functions, surface and volume calculations, and modeling). The test takes approximately 90 min to complete and consists of three sections, each dealing with one of the three videos, which are presented to participants in a random order. In each unit, participants first receive context information about the lesson they are about to watch, the class, the mathematical topic, and the topics covered previously. They can then watch the video vignette once only to simulate a real classroom situation as closely as possible. After watching, participants respond to open-response and rating-scale items that test their perceptual, interpretive, and decision-making skills from a mathematics pedagogical and general pedagogical perspective. Open-response items consist of a prompt that can be freely answered in a text box, while rating-scale items contain a statement to be assessed using a four-point Likert scale, with the options *not correct at all*, *partially incorrect*, *partially correct*, and *fully correct*. In total, the instrument consists of 77 items. The distribution of items per noticing facet and subject-related perspectives is shown in Table 3.1. The small number of

[1] For further details concerning the instrument and its development see Kaiser et al. (2015) and Hoth et al. (2016).

items testing decision-making is due to the complexity of developing those kinds of items and contextual requirements. Perception items are well represented in the general pedagogical item pool, while decision-making items deal particularly with mathematics pedagogical topics. Interpretation items are evenly distributed. These distributions are the result of test construction and data scaling. Item examples can be found in the papers in Chapters 4, 5, and 6 as well as in the descriptions of Kaiser et al. (2015).

Table 3.1 Number of items by noticing facet, subject-related perspective, and item format

Noticing facet	Subject-related perspective		Total
	MPID	PPID	
Perception	3	21	24
Interpretation	25	17	42
Decision-making	8	3	11
Total	36	41	77

Note. MPID—mathematics instruction: perception, interpretation, decision-making, PPID—pedagogy: perception, interpretation, decision-making. The items for subject-relative perspective and item format each add up to the total number of items

3.2.2 Quality of the Instrument

In addition to thorough, well-documented test development in accordance with scientific standards and ensuring the objectivity and reliability of the measurement, a high-quality instrument means validating the test use and test score interpretation (American Educational Research Association, American Psychological Association, National Council on Measurement in Education [AERA, APA & NCME], 2014; Kane, 2013). It is important to note that the validity of a test instrument cannot be demonstrated; only evidence for specific test uses or test score interpretations can be provided (Kane, 2013). In his argument-based approach to validation, Kane (2013) suggested structuring evidence in validation processes in an argument-oriented way and presenting evidence to support inferences about the processes of scoring, generalization, and extrapolation as central steps in test score interpretation. Scoring involves the generation of test scores based on observed test performance. Generalization inference takes into account the representability of a test and the observations used to arrive at a generalized expected test score, for example, based on a statistical model. The extrapolation step includes abstracting from the estimated test performance to a score for the abilities in the entire construct under study or in the domain of interest. Kane (2013) argued that evidence of the plausibility of these three steps must be provided to ensure the validity of test score interpretation. The standards for educational and psychological testing by AERA, APA, and NCME (2014) distinguish five sources of evidence in this context: evidence based on test content, evidence based on response processes, evidence based on internal structure, evidence based on relations to other variables, and evidence for the validity and consequences of testing. In this section, I present evidence for the validity of test use and test score interpretation for the TEDS noticing instrument in the context of the three core publications of this thesis, following Kane's (2013) three steps and adopting the sources of evidence described above.

In the context of my thesis, following the framework presented in Sect. 2.3.2, the observed performance on the instrument's items for measuring teacher noticing was used to create general scores for teacher noticing skills in the facets of perception, interpretation, and decision-making from mathematics pedagogical and general pedagogical perspectives of pre-service and in-service secondary mathematics teachers. The scores were then used to describe the participants' skills and to contrast and compare groups of pre-service and in-service teachers (Publications I and II). Furthermore, I used pre-service teachers' scores in a design with two measurement points to examine changes in their teacher noticing skills over the course of teaching internships (Publication III).

For these test uses, the scoring inference was informed by extensive expert reviews with experienced teachers working in teachers' professional development or the induction phase and academic experts in the field to obtain a scoring rubric for the rating-scale items (Hoth et al., 2016). In addition, an expert-assisted development of a comprehensive coding manual with extensive coding rules and example responses for scoring open-response items was conducted (Kaiser et al., 2015). Objectivity in the coding of the open-response items was achieved by ensuring intercoder agreement for 20% of the items for each investigated sample, using Cohen's κ (Cohen, 1960). Values are reported in the respective studies.

Evidence for the generalizability of the observed test scores was gathered through several steps in the development of the instrument and its empirical investigation. First, in development, the project team created video scripts of prototypical, authentic classroom situations, including lessons on important topics in mathematics secondary education, based on curriculum analyses. These scripts were then reviewed in collaboration with academic experts and experienced teachers (Kaiser et al., 2015). The item creation and selection covered a wide range of important topics from mathematics pedagogy and general pedagogy for high-quality mathematics education (Bastian et al., 2023; Kaiser et al., 2015). Empirically, confirmatory factor analyses established the assumed, theoretically derived structure of the competence and teacher noticing framework used for samples of pre-service and in-service teachers (Bastian et al., 2022; Blömeke et al., 2016; Blömeke, König, et al., 2014; König et al., 2014; Weyers, König, Rott, et al., 2023). Blömeke and König et al. (2015) demonstrated the independence of the test scores from the test's video vignettes. The application of Rasch models in the context of IRT ensured psychometric testing and the investigation of sampling errors, and allowed the estimation of generalized scores (Kane, 2013; Rasch, 1960). The calculated Rasch models showed at least acceptable model fit and reliability for samples of pre-service and in-service teachers (Bastian et al., 2022; Blömeke et al., 2022; Blömeke & Kaiser, 2014; Weyers, König, Rott, et al., 2023).

To support the exploration inference, the item creation and selection was conducted with experienced teachers and based on curricular analyses to ensure good representability of the items used (Kaiser et al., 2015). In addition, the test development and the item creation were strongly informed by the framework of competence and the definition of teacher noticing and its facets. Each item was developed with a certain facet of teacher noticing, a mathematics pedagogical or general pedagogical perspective and corresponding professional knowledge, and a distinct event of the classroom situation in mind (Bastian et al., 2023; Kaiser et al., 2015). Empirical evidence for the exploration warrant and the relationships with other variables were also gathered. Blömeke, König, et al. (2014) and König et al. (2014) demonstrated significant effects of MCK, MPCK, and GPK on teacher noticing and its facets. Subsequent studies have enriched these findings by examining the overall impact model, that is, the relationships between teacher competence, instructional quality, and students' learning gains (see Fig. 2.1 for the model). Blömeke et al. (2022) found a moderate to strong mediating effect of teachers' situation-specific skills and instructional quality in the relationship between teachers' knowledge and students' achievements. Further, they identified teacher noticing skills as having the strongest impact on students' achievements for the competence facets investigated, likely due to the proximal nature of teacher noticing to actual classroom performance. The findings described above provide evidence for the hypothesized structure of teachers' professional competence and the role of teacher noticing in it, as well as the impact of teacher noticing on student variables (see Fig. 2.1 and 4.1).They thus establish the investigated TEDS teacher noticing measure as capturing a situation-specific part of teachers' competence and ensure the validity of test use and test score interpretation.

However, correlations between the three facets of teacher noticing and facets of professional knowledge have not yet been published as part of the TEDS research program. Since these provide quick insight into the relationship between teacher noticing and other variables, I included them here as additional support for the extrapolation inference. Relationships between situation-specific and more cognitive-oriented skills are of particular interest, as they demonstrate the interaction between situation and cognition (Kaiser et al., 2017). Thus, I investigated the correlations between perception, interpretation, and decision-making as situation-specific skills with MCK, MPCK, and GPK, drawing data from $n = 201$ participants in the TEDS-I and TEDS-V studies who participated in the TEDS noticing survey and the MCK, MPCK, and GPK tests. Detailed information on the structure, items, and development of the three knowledge tests can be found in Tatto (2013), Döhrmann et al. (2010), and Blömeke and König (2010a). The participants were in-service teachers from four German federal states (Hamburg, Hesse, Thuringia, and Saxony). They were 56% female, had a mean age of 42.4 years (SD = 10.8 years), and had an average of 14.6 years of teaching experience (SD = 11.1 years). To estimate the latent, attenuation-corrected correlations of all six facets of teachers' professional competence, I computed a six-dimensional Rasch model using weighted likelihood estimators (WLEs) and the software *ConQuest 5.33.5* (Adams et al., 1997–2023). Reliabilities of the estimators and the correlations are shown in Table 3.2. To promote comparability with other studies, manifest and latent attenuation-corrected correlations are presented. As latent correlation are adjusted for measurement errors, they are usually higher than manifest correlations. Note that in Table 3.2 the latent attenuation-corrected correlations are reported below the diagonal, while the manifest empirical Pearson correlations are reported above the diagonal.

All scales showed at least acceptable reliability, with a somewhat lower value for decision-making, which may be due to the low number of decision-making items in combination with a complex construct and the variety of decision-making contexts represented in the instrument. The correlations between the facets of teacher noticing and the components of professional knowledge among each other were comparable to results in previous studies in the TEDS research program, with a somewhat higher correlation between MCK and MPCK (Bastian et al., 2022; Blömeke, Houang, & Suhl, 2014; Blömeke et al., 2022; Blömeke & Kaiser, 2014). GPK was also strongly related to domain-specific MCK and MPCK, as shown by latent correlations of $r = 0.78$ and $r = 0.81$. Correlations were

Table 3.2 Reliabilities, Manifest and Attenuation-corrected Correlations for Facets of Teacher Noticing and Professional Knowledge

Variable		WLE reliability	(1)	(2)	(3)	(4)	(5)	(6)
(1)	Perception	0.58	–	0.57	0.40	0.15	0.19	0.16
(2)	Interpretation	0.73	0.87	–	0.51	0.26	0.25	0.20
(3)	Decision-making	0.45	0.83	0.90	–	0.21	0.22	0.20
(4)	MCK	0.85	0.37	0.41	0.19	–	0.87	0.75
(5)	MPCK	0.85	0.40	0.41	0.20	0.96	–	0.77
(6)	GPK	0.91	0.39	0.34	0.32	0.78	0.81	–

Note. Correlations above the diagonal are manifest; those below the diagonal are attenuation-corrected. MCK—mathematics content knowledge, MPCK—mathematics pedagogical content knowledge, GPK—general pedagogical knowledge

expected, as these three facets are all conceptualized as belonging to the domain of professional knowledge. The high magnitude of the relationship between the three knowledge facets may be a sign of expertise in teaching for this sample of experienced in-service teachers, as experts tend to have more strongly correlated and connected knowledge than novices (Berliner, 2004; Bromme, 1992; Krauss et al., 2008).

The correlation between the facets of teacher knowledge and teacher noticing was of particular interest. The correlations ranged from $r = 0.19$ to $r = 0.41$, indicating small to moderate effects of the relationships (Cohen, 1988). The correlation patterns of MCK, MPCK, and GPK to the noticing facets were somewhat comparable, which is consistent with the strong correlation between the three knowledge facets. For MCK and MPCK, the correlation with interpreting was higher than for the other two facets. This was expected, since teacher noticing is a knowledge-based construct, and interpreting is the noticing facet that specifically links cognitions to situations (Kaiser et al., 2015; M. G. Sherin, 2007; van Es & Sherin, 2002). For GPK, the correlation with perception was highest, which may imply that GPK is particularly relevant for identifying important events in classroom situations but could also be an effect of the number of perception items related to the general pedagogical perspective.

3.3 Research Questions and Structure of the Thesis

In light of the overall research interest and the research subjects A to G described in Sect. 1.1, and based on the current state of the art as well as the identified research gaps, this thesis addresses the following research questions through three distinct core publications.

Publication I (research subjects A and B).

(1) Can the three distinguished noticing facets of perception, interpretation, and decision-making be separated empirically using quantitative models and measured through reliable scales? If yes, how are the scales representing the facets intercorrelated in this model?
(2) How do teacher noticing and its facets differ between different pre- and in-service teacher groups with varying degrees of professional teaching experiences?

Publication II (research subjects C and D).

(3) What differences exist between the teacher noticing of groups with different levels of teaching experience in managing the cognitive demands of teacher noticing measures and related teaching situations?

Publication III (research subjects E, F, and G).

(4) To what extent do the facets of teacher noticing at the beginning of teaching internships condition change over the course of the internships, considering autoregressive and cross-lagged effects?
(5) To what extent do teaching internships affect changes in teacher noticing skills?
(6) To what extent do certain teaching internship process variables (i.e., learning time, teaching practice activities, and mentor support) influence teacher noticing skills?

Publication I

4

The first publication of this thesis, "Teacher noticing and its growth toward expertise: An expert–novice comparison with pre-service and in-service secondary mathematics teachers," was published in 2022 in the journal Educational Studies in Mathematics. It addressed the lack of studies focusing on the development of teacher noticing as part of teacher expertise and dealt with the facet structure of teacher noticing and expertise- and teaching experience-relating differences in the noticing of three groups of pre-service and in-service teachers. In reference to the expanded impact model of teacher education and competence introduced in Section 2.1 (see Fig. 2.1), the paper was located between area (1) "Teacher education and professional development" and area (2) "Teacher competence" and connected different stages of teacher education and levels of teaching experience with the situation-specific skills of teacher competence (see an excerpt of the expanded impact model in Fig. 4.1). As the results were also compared and connected to similar studies from other cultural contexts, area (5) "Context" was also affected (Fig. 4.1).

The paper focused on whether the three facets of teacher noticing—perception, interpretation, and decision-making—can be separated empirically and how these facets are correlated. Further, differences in the noticing among the three groups of pre-service and in-service teachers with different lengths of teaching experience were examined. To investigate these questions, the authors analyzed a sample of 457 individuals who participated in the TEDS teacher noticing survey in the TEDS-FU, TEDS-I, TEDS-V, or TEDS-V-T studies, and divided this sample into three groups: master's students, early career teachers, and experienced teachers. Analyses comprised the comparison of different Rasch and

A. Bastian, *Teacher Noticing of Pre-service and In-service Secondary Mathematics Teachers*, Perspektiven der Mathematikdidaktik,
https://doi.org/10.1007/978-3-658-45619-1_4

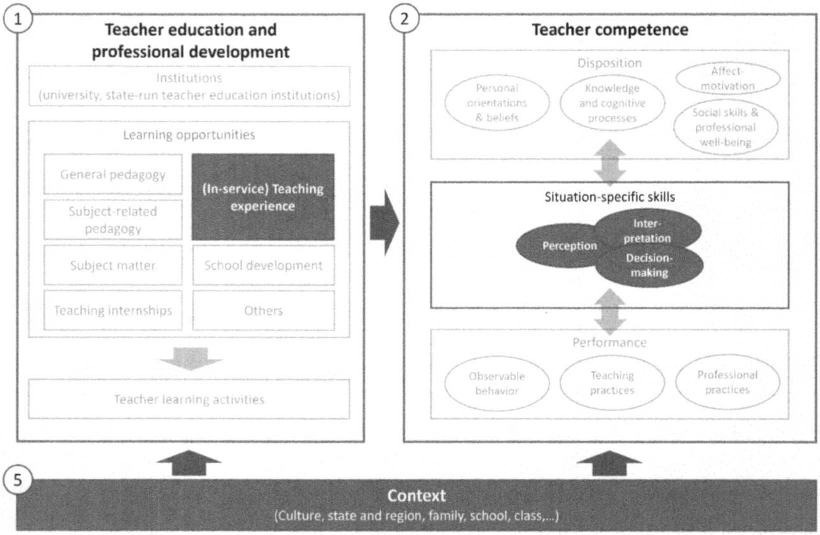

Note. **Areas examined by the research presented in Publication I are marked in blue.**

Fig. 4.1 Excerpt of the Expanded Impact Model to Illustrate the Contribution of Publication I

structural equation models, as well as variance and regression analyses. The results demonstrated that the three-dimensional model of perception, interpretation, and decision-making fit better than a one-factor model of teacher noticing, supporting the conceptual and empirical distinction of the three facets. In addition, the authors provided evidence for an increase in teacher noticing skills with increasing teaching experience, as in-service teachers scored significantly higher than pre-service teachers. However, since early career teachers and experienced teachers did not differ significantly in perception and interpretation, and early career teachers outperformed experienced teachers in decision-making, the findings also indicated stagnation effects in teacher noticing skills for in-service teachers. Additional regression analyses even showed small but significant declines with years of teaching experience for the group of in-service teachers in all three noticing facets. Because these findings contradict comparable studies in other cultural contexts but are consistent with studies in a German context, this publication emphasized the cultural dependence of teacher noticing skills. The unexpected differences between early career and experienced teachers called for further investigation, which is part of Publication II.

4.1 Individual Contribution Statement

Gabriele Kaiser, Johannes König, Dennis Sommer (née Meyer), Björn Schwarz, and I collaborated on the research project reported in Publication I. In previous studies in the TEDS research program, the three facets of teacher noticing had not been empirically separated and investigated, as teacher noticing was divided into mathematics pedagogical and general pedagogical perspectives; thus, interpretation and decision-making had to be combined due to the low number of items (Blömeke, König, et al., 2014; König et al., 2014; Yang, König, & Kaiser, 2021). I proposed combining both perspectives to enable the investigation of all three facets of the theoretical framework separately. Gabriele Kaiser, Johannes König, Dennis Sommer, and I planned the publication together and selected the appropriate sample for the analyses. Dennis Sommer did the data compilation and cleaning from the four different data sets, as well as the scaling itself. I conducted all other analyses—the comparisons of the three groups of teachers, the variance and regression analyses, and the structural equation model analyses—and analyzed all results, including the IRT scaling. In addition, Björn Schwarz and I performed in-depth qualitative item analyses to complement the quantitative results. These were later excluded in the revision of the paper based on reviewers' comments but were a basis for the item analyses conducted in Publication II. I wrote the introduction, methods, results, and discussion myself, and compiled all tables and figures. Gabriele Kaiser and I collaborated on the literature review. Subsequently, I revised the paper based on feedback from all co-authors. Gabriele Kaiser submitted the manuscript to the journal. All revisions based on the three rounds of peer review were made by me in close collaboration with Gabriele Kaiser and Johannes König.

4.2 Publication Imprint

The following manuscript was published originally by Springer Nature in the journal Education Studies in Mathematics in 2022[1]. It is reprinted here under the terms of the Creative Commons CC BY license.

Teacher noticing and its growth toward expertise: an expert–novice comparison with pre-service and in-service secondary mathematics teachers

Anton Bastian[1] (ORCID: 0000-0002-1177-6336), Gabriele Kaiser[2] (ORCID: 0000-0002-6239-0169)[1,4*], Dennis Meyer[3] (ORDCID: 0000-0003-1608-124X), Björn Schwarz, Johannes König[4] (ORCID: 0000-0003-3374-9408)

[1]University of Hamburg, Von-Melle-Park 8, 20146 Hamburg, Germany
[2]University of Hamburg, Australian Catholic University, Von-Melle-Park 8, 20146 Hamburg, Germany
[3]University of Hamburg (ProfaLe), Bogenallee 11, 20144 Hamburg, Germany
[4]University of Vechta, Driverstraße 22, 49377 Vechta, Germany
[5]University of Cologne, Gronewaldstraße 2a, 50931 Cologne, Germany
[*]Corresponding author: Gabriele Kaiser, gabriele.kaiser@uni-hamburg.de

Abstract

Although strong references to expertise in different theoretical approaches to teacher noticing have been made in the last decades, empirical knowledge about the development of teacher noticing from novice to expert level is scarce. The present study aims to close this research gap by comparing three different groups of mathematics teachers with different degrees of professional teaching experience—pre-service teachers at the master's level, early career teachers, and experienced teachers—using data sampled in the frame of the research program from the Teacher Education and Development Study in Mathematics (TEDS-M). Furthermore, the construct of teacher noticing is assessed in a differentiated way by analyzing different noticing facets. Findings confirm that three facets of teacher noticing can be empirically distinguished—perception of important

[1] Bastian, A., Kaiser, G., Meyer, D., Schwarz, B., & König, J. (2022). Teacher noticing and its growth toward expertise: An expert–novice comparison with pre-service and in-service secondary mathematics teachers. *Educational Studies in Mathematics*, *110*(2), 205–232. https://doi.org/10.1007/s10649-021-10128-y

classroom events, their interpretation, and decisions regarding further developments. The results reveal a considerable increase in professional noticing between master's students and practicing teachers. However, in contrast to other studies, among examples from East Asia, a stagnation or decrease in professional noticing between early career teachers and experienced teachers could be observed. Overall, the study highlights the cultural dependency of expertise development regarding teachers' noticing.

Keywords: Teacher noticing · Expertise development · Expert–novice comparison · Mathematics education · Pre-service and in-service secondary mathematics teachers

4.2.1 Introduction

During the last two decades, teacher noticing has increased in significance, particularly in mathematics teaching and mathematics teacher education (Schack et al., 2017; Sherin et al., 2011a). Although noticing is a common act, it is very specific to and within the teaching profession. As Mason (2002) notes, "every act of teaching depends on noticing: noticing what children are doing, how they respond, evaluating what is being said or done against expectations and criteria, and considering what might be said or done next" (p. 7). Thus, noticing poses the "primal questions of teaching" (Sherin et al., 2011b, p. 3).

Currently, noticing is considered integral to professional teacher competency and expertise (Kaiser et al., 2017) and seen as an essential component of teachers' expertise. It is the subject of numerous recent empirical studies, particularly in video-based research programs (overviews can be found in Santagata et al., 2021; Stahnke et al., 2016). Nevertheless, few studies have focused on the development of teacher noticing as part of their expertise development. Therefore, this paper aims to describe and analyze the development of pre-service and in-service mathematics teachers' expertise, focusing on teacher noticing by comparing master's students in initial teacher education with early career teachers and experienced teachers in the context of the Teacher Education Development Studies in Mathematics (TEDS-M) research program in Germany (Kaiser & König, 2019).

Despite increasing importance of teacher noticing in the educational discussion, consensus on the conceptualization of teacher noticing as a construct is lacking. While some theoretical approaches conceptualize noticing as a holistic concept (Mason, 2015), an analytical perspective on teacher noticing has gained importance in the last decade, particularly in empirical research distinguishing its different facets (for an overview, see Santagata et al., 2021). Although Sherin et al. (2011b) identify "attending to particular events" and "making sense of events" as two central aspects of noticing, disagreement surrounds the inclusion of a third aspect—"decision-making"—as described in Jacobs et al. (2010) and the conceptualization of our research team (Kaiser et al., 2015). However, not only are certain facets of noticing controversial but so are efforts to expand the focus of its attention beyond an exclusive focus on children's mathematical thinking and their strategies (Jacobs et al., 2010) to encompass whole lessons, including attending to important classroom incidents and decision-making regarding possible continuations of lessons and alternative student-teacher interactions (König et al., under review). Thus, to enrich this discussion with empirical arguments, we use our broadened conceptualization of noticing to examine the measurement of this construct comprising three facets—perception, interpretation, and decision-making—and test its empirical validity.

Although the necessity of teachers' expertise in teaching is emphasized in the current discussion, only a few empirical studies have analyzed the development of teachers' expertise and compared the noticing of novice and expert teachers. For example, early empirical studies of expertise by Berliner (1988) and colleagues (Carter et al., 1988; Sabers et al., 1991) demonstrated that expert teachers noticed issues concerning simultaneity, multidimensionality, and immediacy at higher levels than early career novice teachers or pre-service teachers in initial teacher education. Furthermore, expert teachers focused more on students and important content-related aspects of teaching. A more recent study by Jacobs et al. (2010) comparing various expertise groups of (pre-service) primary teachers indicated that teachers' growth and expertise development were decisive factors for the development all three noticing facets with teaching experience (see also Yang et al., 2021b).

Overall, despite this important work, the question of whether these three noticing facets of perception, interpretation, and decision-making can be separated empirically based on quantitative data from large-scale studies remains open. Furthermore, how these three facets differ empirically across various teachers' groups with varying professional expertise requires further analysis. Finally, the influence of the conceptualization of teacher noticing merits investigation with regard to how teachers' noticing differs across various groups if the focus of noticing is broadened from children's mathematical thinking to activities encompassing entire lessons. The present study aims to close this gap by empirically examining how different pre- and in-service teachers' groups notice, adopting a more analytic perspective by empirically differentiating the noticing construct into three different facets.

4.2.2 Literature Review, Theoretical Framework, and Research Questions

4.2.2.1 Literature Review on Expertise Development in Noticing

Research on teacher noticing has mainly been conducted within the last two decades and has gained increased attention due to its focus on students and their mathematical thinking and its relevance to quality-oriented teaching. Despite the consensus regarding the central characteristics and aspects of teachers' noticing, the construct has largely been conceptualized heterogeneously (Dindyal et al., 2021; Schack et al., 2017; Sherin et al., 2011a; Stahnke et al., 2016). Given the extensive body of research on teacher noticing, this review includes work that directly relates to the study described in this paper. First, theoretical perspectives on teacher noticing and its conceptualizations and foci are displayed, followed by research on the development of teacher noticing.

As previously mentioned, the current discussion on teacher noticing can be characterized by a certain inconsistency concerning the theoretical orientation and conceptualization of teacher noticing. To present a systematized view on the current discussion, we refer to a classification proposed in a systematic literature survey on noticing education among mathematics teachers with a focus on video-based programs (Santagata et al., 2021). Santagata et al. (2021) described four theoretical perspectives on teacher noticing.

The first perspective—called the cognitive-psychological perspective—characterizes noticing as the underlying, subconscious mental processes in which teachers engage during teaching or teaching observation. In their seminal paper on the promotion of teacher noticing by video clubs, van Es & Sherin (2002) distinguished the following three facets of teacher noticing:

> (a) identifying what is important or noteworthy about a classroom situation; (b) making connections between the specifics of classroom interactions and the broader principles of teaching and learning they represent; and (c) using what one knows about the context to reason about classroom interactions. (p. 573)

This approach has been further developed by several scholars. For example, Jacobs et al. (2010) refocused the first two facets of noticing as "attending to children's strategies" and "interpreting children's mathematical understandings" and included a unique third facet of noticing, "deciding how to respond on the basis of children's understanding" (pp. 172–173). A similar conceptualization has been developed within the TEDS-M research program, in which decision-making as the third facet of noticing plays an important role. However, in contrast to Jacobs et al. (2010), the focus of the conceptualization of the TEDS-M research program is broadened by focusing on the whole lesson under mathematics educational and pedagogical perspectives. This construct will be elaborated in the theoretical framework section (Kaiser et al., 2015; Yang et al., 2021a). The role played by specific mathematical content domains such as fractions, ratios, multiple representations, or functions is only emphasized in a few studies (see, e.g., Dreher & Kuntze, 2015; Friesen & Kuntze, 2021; Ivars et al., 2020). Another stance is taken by Stockero and Van Zoest (2013) and Stockero (2021), who foreground "in-the-moment noticing" by zooming in on pivotal teaching moments or interruptions in the flow of lessons, which indicate opportunities to modify instructions in order to enhance student mathematical understanding. In their most recent paper, van Es & Sherin (2021) further developed their framework by rephrasing and regrouping these three facets as "attending" and "interpreting." They included a new third facet—shaping—defined as the "act of creating interactions that provide increased opportunities to attend to and interpret noteworthy mathematical interactions" (p. 17); this new facet reinforces a knowledge-driven perspective on noticing.

In the second perspective, noticing is conceptualized as situated and socially constructed. This socio-cultural perspective draws on the seminal work of Goodwin (1994) on professional vision and is intertwined with the first perspective. Goodwin (1994) described professional vision as "socially organized ways of seeing and understanding events that are answerable to the distinctive interests of a particular social group" (p. 606), emphasizing the socio-cultural aspects of professional ways of seeing in certain professions (e.g., law or archaeology). Although the term "professional vision" has been adopted in several approaches and studies, its usage does not always involve socio-cultural framing, and some instead employ a cognitive-psychological perspective that also takes a quality-oriented focus on noticing (e.g., Seidel & Stürmer, 2014). More recently, theoretical conceptualizations emphasizing power and equity have been developed. These are more strongly connected with the original understanding of the professional vision of Goodwin. These conceptualizations challenge the focus on cognition by emphasizing the material and reciprocal nature of noticing (Dominguez, 2019) or introducing a sociopolitical perspective in which anti-deficit noticing shall be developed (e.g., Louie, 2018; Louie et al., 2021).

A third theoretical perspective on noticing—discipline-specific—was developed by Mason (2011) and conceptualizes teacher noticing as a "collection of practices designed to sensitize oneself so as to notice opportunities in the future in which to act freshly rather than automatically out of habit" (p. 35). Although this conceptualization of noticing attends to mental processes, differences in the cognitive-psychological perspective can be identified as the construct of sensitized awareness of the teachers is put in the foreground, which should "be methodical without being mechanical" (Mason, 2002, p. 61). The perspective by Mason (2002) is characterized by its focus on the professional development of teachers, for which Mason (2002) developed a set of practices, including systematic reflection and recognizing, which makes this approach different than the cognitive-psychological perspective.

The fourth perspective corresponds to the expert–novice paradigm developed by Berliner (1988), which can be described as a kind of precursor to research on teacher noticing (Lachner et al., 2016), although the noticing construct is not explicitly used in this perspective. The approach focuses similarly on the cognitive-psychological perspective on teacher ability to attend to important elements of teaching events, create coherent interpretations, and ignore distracting elements (Sabers et al., 1991). Overall, the processes of interpreting classroom situations, making sense of important events, and developing connections are described in this perspective as individual cognitive mental processes carried out by teachers, revealing connections to the discourse of teacher noticing. This perspective will be revisited during the discussion of the development of teachers noticing.

More recent work aims to overcome these different normative orientations (e.g., Scheiner, 2021); however, it thus far remains open just how strongly these new approaches will influence the discussion.

Concerning the relevance of these four theoretical perspectives, Santagata et al. (2021) report in their systematic literature survey that most studies examined (about 90%) referred to the cognitive-psychological perspective using the term "noticing," while the other three perspectives played more minor roles. In line with the dominance of the cognitive-psychological perspective is a more analytical view on noticing that advocates distinguishing different facets of noticing. As Santagata et al. (2021) point out, most examined studies in the systematic literature survey focus on the facets of attending/perceiving and interpreting/reasoning (about 80%), while fewer studies (about 37%) included responding/

decision-making. Concerning the object of the study, most of the examined studies analyzed teacher noticing of student thinking (83%), followed by a focus on instructional practices and classroom discourse (57%). Mathematical topics were addressed in fewer studies (20%). However, despite the dominance of the analytic approach, which is separating the different facets of noticing, empirical evaluation and analysis is often based on a holistic approach concerning the facets of noticing; that is, even if facets are distinguished, they are not assessed separately. Only a few instruments have measured the facets of noticing separately in compliance with high psychometrical standards, namely the observer/observer extended tool (Seidel & Stürmer, 2014), pre-service teachers' professional vision of inclusive classrooms (PVIC) (Keppens et al., 2019), the video-vignettes developed within the TEDS-M research program (Kaiser et al., 2015), and the instrument by Steffensky et al. (2015).

Another stance is taken in a recent literature survey by Amador et al. (2021), who analyzed methodological approaches to support and analyze pre-service teacher noticing by focusing on the two most currently influential theoretical frameworks on noticing development within the US discussion. The first framework by van Es (2011) on learning to notice mathematical thinking distinguished four levels of expertise in noticing, providing descriptors for each. The second framework, developed by Jacobs et al. (2010) on the professional noticing of children's mathematical thinking, empirically described the development of teacher noticing across different expertise groups using limited, lacking, or robust evidence as indicators for noticing. The results of the literature survey by Amador et al. (2021) indicated that more studies using the learning to notice framework reported positive results regarding progress in the development of noticing compared to studies using the professional noticing framework. However, as the effects reported in the included studies cover different participant groups and different designs, the question of how teacher noticing develops empirically remains open.

In the following, the research on the development of teacher noticing is delineated, and the fourth theoretical perspective is described in more detail. As alluded, earlier research on expertise and expert–novice comparisons identified important results regarding the development of expertise in noticing, which must be taken up in current discussions, although the construct of noticing was not explicitly used in these studies. In his seminal work, Berliner (1988) distinguished five different stages in teacher expertise development—novices, advanced beginners, competent teachers, proficient teachers, and expert teachers. While novices are characterized by the need for context-free rules and inflexible teaching and competent teachers act rationally but only with minor flexibility, expert teacher behavior can be described as "arational" (Berliner, 1988, p. 5), which is fluid, holistically, and unconsciously flexible. Experience plays a crucial role in expertise development and should be regarded as a necessary yet insufficient condition for expertise (Berliner, 2004; Palmer et al., 2005). In empirical studies comparing novice and expert teachers, Berliner (1988) highlighted the difficulties that novices face in interpreting classroom phenomena. Novice teachers were unable to predict teaching and student behavior, particularly concerning student errors. Further work from this research group shows a strengthened relationship to the noticing discussion. For example, the study by Sabers et al. (1991) described the differences between novice and expert teachers regarding the perception and interpretation of important classroom events.

These results are confirmed more recently by the literature survey by Stahnke et al. (2016), which suggests that novices tend to have difficulties perceiving or interpreting student work, which apparently can be improved by video-based professional development programs. Decision-making appears to be the most ambitious facet of noticing for novices, as pre-service teachers seem to have difficulties understanding student mathematical thinking and their solution processes in a constructivist sense.

Nonetheless, a key area of concern is that although expertise is an important reference point in the theoretical frameworks of many studies on noticing, most current studies are restricted to pre-service teachers and do not include practicing teachers, which means that limited knowledge is available concerning the development of teacher expertise in noticing, as was noted in the systematic literature survey by Santagata et al. (2021). In a more extensive literature survey focusing on teacher noticing, König et al. (under review) indicated that only a few of the papers reviewed included both pre-service and in-service teachers and allowed examination of the development of teacher noticing, either in a cross-sectional or a longitudinal way, although the researchers used different constructs and definitions of novice and expert. Most of these few papers offered quantitative analyses of the differences between novice and expert teachers, mainly described as pre-service and in-service teachers. Only these few papers included in the literature survey focused explicitly on and reported noticing progress between novice and expert teachers. The subjects covered ranged from mathematics education (Jacobs et al., 2010) to elementary science (Meschede et al., 2017) or physical education (Reuker, 2017) or focused on classroom management (Gold & Holodynski, 2017). The studies clearly indicate that experts outperform novices in professional noticing.

One of the rare exceptions concerning comparisons of experts and novices is the seminal study by Jacobs et al. (2010) that provided insight into expertise development. They compared four different expertise groups ranging from pre-service elementary school teachers and three groups of practicing elementary school teachers that differed in terms of their years of teaching experience and were at different stages of a professional development program concerning children's mathematical thinking. Video clips and written student work were used to elicit teacher noticing of student mathematical thinking. The participants answered prompts concerning the three identified facets of noticing, attending to children's strategies, interpreting children's understanding, and deciding how to respond based on children's understanding. Within the study, an extensive coding manual with indicators for the adequacy of the analysis was developed using the categories of robust, limited, or lack of evidence. Jacobs et al. (2010) primarily identified a monotonic development across the four groups for all three noticing facets, "indicating that increased experience with children's thinking was related to increased engagement with children's thinking on the professional-noticing tasks" (p. 181). Their findings suggest that professional teaching experience and development programs facilitate expertise development in attending to children's strategies and interpreting children's understanding; however, they could not substantiate the same hypothesis for deciding how to respond. They also reported that "professional development seems to provide support for developing expertise in all three component skills" (p. 182), which provides evidence for the effectiveness of continued long-term professional development in this area, especially regarding leadership activities.

The results are confirmed by Yang et al. (2021b), who compared Chinese pre-service and two groups of in-service mathematics teachers with different degrees of teaching experience with respect to growth in their professional noticing. Although the study differentiated three facets of noticing, the facets interpretation and decision-making are joined in the results due to the low number of items on decision-making. The study reports mean differences in noticing across the three expertise groups that can be interpreted as a "nearly linear growth…with significant differences identified between pre-service and experienced teachers and only small differences between pre-service and early career teachers" (p. 29). Furthermore, the results using differential item functioning (DIF) analyses point out that pre-service and early career teachers showed strengths in more reform-oriented Western topics, such as cooperative learning or mathematical modeling, which they may have learned during their university study, whereas experienced teachers showed strengths in analyzing student mathematical thinking. They found that the three noticing facets develop differently between the three expertise groups, with perception being better developed at the pre-service and beginning stages of teaching in contrast to interpretation and decision-making (which are more difficult to develop), which is aligned with the results by Jacobs et al. (2010).

Further results regarding expert–novice comparisons of teacher noticing are reported by other studies. For example, Dreher and Kuntze (2015) indicated that in-service teachers more strongly appreciated the usage of multiple representations and the role of changes of representations for fostering student understanding than pre-service teachers. Huang and Li (2012) emphasized qualitative differences between early career teachers and highly experienced teachers concerning the development of mathematical thinking and knowledge, student participation, and teacher motivation, with more experienced teachers attaining higher achievements.

Overall, more empirical research is needed to better understand the development of teacher noticing and the factors influencing this development, especially the role of teaching experience and expertise, its relation to the development of the different noticing facets, and how these factors manifest in a Western context.

4.2.2.2 Context of the Study and Theoretical Framework

As previously explained, in the last decade, the paradigmatic differences between cognitively oriented frameworks on teacher competence that emphasize teacher knowledge and situated approaches putting social and situated aspects of teaching in the foreground have been challenged. In order to overcome these paradigmatic differences between cognitive and situated frameworks, Blömeke et al. (2015) developed the new theoretical approach of competence as continuum, which has become a prominent framework for teacher competencies. This approach called for situation-specific skills mediating between teacher knowledge and beliefs as dispositions and teacher performance.

The current study, which is carried out within the TEDS-M research program, departs from this discussion around these paradigmatic differences. To facilitate a better understanding of the context of the current study, this section briefly describes the various studies carried out within this program (for details of the development, see Kaiser et al. (2017)). The TEDS-M research program was initiated by the international TEDS-M study 2008, which took place across 17 countries from 2008 to 2010 and evaluated the professional knowledge of pre-service teachers for primary level and mathematics teachers for secondary level at the end of their study (Tatto et al., 2012). TEDS-M can be characterized as a cognitively oriented study with a focus on the professional knowledge of pre-service teachers referring to the well-known classification of teacher knowledge by Shulman (1986). Within the study, mathematical content knowledge (MCK), mathematics pedagogical content knowledge (MPCK), and general pedagogical knowledge (GPK) in a sub-sample of countries, including Germany, were evaluated; these knowledge domains were complemented by affective-motivational characteristics, such as beliefs about mathematics and its teaching and epistemological beliefs. The standardized testing instruments consisted mainly of multiple-choice items and a few open-ended items and were carried out as paper-and-pencil tests. The evaluation presented strong country differences, with the participating pre-service teachers from East Asia achieving the best results and those from South America and Africa receiving the lowest results in all knowledge facets. Further analyses indicated, among others, the strong influence of opportunities to learn and background variables (Blömeke et al., 2012).

A follow-up study of TEDS-M (TEDS-FU) by German scholars further developed the conceptualization and the instruments of TEDS-M towards a situated approach including situation-specific skills—called teacher noticing—to compare the development of teacher expertise from pre-service teachers to early career teachers with instruments that more closely resembled actual classroom practices. Theoretically, the cognitively oriented knowledge framework was enriched by additional situated facets covering perception, interpretation, and decision-making as facets of noticing (the so-called PID model). For the measurement of these newly developed noticing facets, a standardized measurement instrument was developed based on several short video-vignettes and extensive coding manuals using closed and open-ended items. The study was conducted from 2011 to 2014, with voluntary participants from the original sample of TEDS-M (for a description, see Kaiser et al. (2015)). Among others, the study analyzed the structure of mathematics teacher competence, distinguishing stable teacher cognitions consisting of subject-related knowledge facets (MCK and MPCK) with a facet of fast error recognition and situation-specific cognitions with general pedagogical knowledge (GPK) and noticing (PID) (Blömeke et al., 2016). Concerning the development of mathematical and mathematics pedagogical content knowledge, the study noted a significant decrease of MCK, which is strongly influenced by MCK measured at the end of teacher education, and no change of the knowledge level of MPCK with less influence of MPCK measured at the end of teacher education. These results point to the strong impact of school practice on the development of MPCK, which is aligned with results by the expertise research (Berliner, 1994). However, the amount of school practice appears inconclusive, which points to the relevance of a "deliberate practice" (Ericsson et al., 1993).

The theoretical framework and the instruments of TEDS-FU have been further enriched, among others by a situation-specific test on classroom management expertise (König, 2015) and instruments for the in vivo evaluation of instructional quality (Schlesinger et al., 2018). These instruments were used in further studies, namely TEDS-Instruct and TEDS-Validate, which included student learning gains within the entire impact model of the TEDS-M research program. The analysis of the link between pedagogical knowledge and situation-specific classroom management expertise, instructional quality, and secondary student mathematics achievement indicated influences of teacher competence on instructional quality and student mathematical progress (König et al., 2021). Further analyses concerning this important impact model are under review.

Finally, within the TEDS-East-West study, most of the instruments were transferred to an East Asian context, evaluating, among others, the relation between teacher knowledge and their professional noticing. Yang et al. (2021a) reported a stronger relation between teacher knowledge and interpretation/decision-making facets of noticing compared to perception. In another study (Yang et al., 2021b), the development of teacher noticing across different expertise groups has been explored, reporting a nearly linear development over the different expertise groups. This is aligned with the results by Jacobs et al. (2010). This will be revisited in the discussion section.

In the following, our own framework on teacher noticing is described, which was developed within the discourse outlined above. We conceptualize noticing as a set of situation-specific skills comprising three interconnected facets: "(a) perceiving particular events in an instructional setting; (b) interpreting the perceived activities in the instructional setting and; (c) decision-making, either as anticipating responses to students' activities or as proposing alternative instructional strategies" (Kaiser et al., 2015, p. 374); this construct is termed the "PID model" (Blömeke et al., 2015). This conceptualization of noticing has specific characteristics that make it distinct from other conceptions, such as the original noticing approach developed by van Es & Sherin (2002), the professional vision approach by Seidel and Stürmer (2014), and further developments described in the survey paper by Dindyal et al. (2021).

This distinct theoretical framework is deliberately not referring to the construct of attending, which is the typical terminology in the noticing discussion as described above, as this construct is strongly related to interpreting; both facets are described as cyclical process of perception and interpretation within this discussion (Sherin et al., 2011b). Departing from the expert–novice paradigm in which the construct perception is widely used to describe the first phase of teacher actions in an instructional setting and by restricting the construct to observable, discernable incidents (Berliner, 2001; Carter et al., 1988), we conceptualize perception more narrowly as "seeing something without or with only minor reference to interpretation." Apart from this different conceptualization of attending and interpreting in our framework, we include decision-making as a third facet of noticing and an indispensable part of noticing as teacher actions in classrooms are essential to their expertise (Erickson, 2011).

Our framework moves forward the focal point of many studies, which delineate as core of their study the focus on student mathematical thinking, strategies, or reasoning (e.g., Jacobs et al., 2010 or Sherin & van Es, 2009) or in-the-moment noticing (Stockero, 2021; Stockero & Van Zoest, 2013). Decisive for our broadened understanding of noticing is the connection between noticing and quality-oriented mathematics education with a clear content-related orientation besides the pedagogical perspective (similarly to Ivars et al., 2020). In detail, we consider various aspects as decisive for a comprehensive understanding of teacher noticing, among others the various dimensions of instructional quality, namely classroom management, potential for students' cognitive activation, individual learning support, and scaffolding, considering both pedagogical and mathematical perspectives as implemented in many empirical studies (Praetorius et al., 2018). Furthermore, in our framework, a broad range of classroom incidents is combined with aspects of instructional quality, such as preventing classroom disturbances, addressing heterogeneous groups of students, and designing individualized and collective mathematical teaching and learning trajectories. Finally, possibilities of continuing a lesson or developing a new teaching sequences are included considering among others subject-related misconceptions or errors of students, and possible alternative of teacherstudents' interactions (Yang et al., 2021a).

With this comprehensive understanding of noticing, we model a broad range of teachers' situation-specific skills and, with reference to Blömeke et al. (2015), integrate teacher noticing as a complement to the former cognitive-oriented and knowledge-based framework of teacher competence, which characterizes teacher competence within the scope of the TEDS-M research program.

Shaped by our theoretical framework of noticing and based on the shortcomings in empirical studies on teacher noticing identified above, our study addresses the following research questions:

1. Can the three distinguished noticing facets of perception, interpretation, and decision-making be separated empirically using quantitative models and measured through reliable scales? If yes, how are the scales representing the facets intercorrelated in this model?
2. How do teacher noticing and its facets differ between different pre- and in-service teacher groups with varying degrees of professional teaching experiences?

4.2.3 Methodology

Due to the complexity of our research questions, we draw our data from several studies conducted between 2011 and 2020 that were embedded in the TEDS-M research program to obtain a rich sample of cross-sectional data.

4.2.3.1 Sample and Participants

The sample consists of three groups of pre-service and in-service secondary level mathematics teachers (N = 457) who participated in a survey of one of the following studies of the TEDS-M research program:

- TEDS-FU: 2011–2014, early career teachers from all over Germany, former participants of the study TEDS-M
- TEDS-Instruct (TEDS-I): 2014–2016, practicing teachers from the federal state of Hamburg, Germany
- TEDS-Validate (TEDS-V): 2016–2019, practicing teachers from three federal states in Germany (Thuringia, Saxony, Hesse)
- TEDS-Validate-Transfer (TEDS-V-T): 2020–ongoing; students pursuing master's degrees in education from six German universities

To ensure that the groups were clearly distinguished by their years of teaching expertise, the participants were divided into three groups based on their teaching experience while also taking study affiliation into account. The master's students from TEDS-Validate-Transfer ($n = 110$) had no professional teaching experience[2] and, hence, are novices according to the stage model by Berliner (1988). Participants with about 4.6 years of teaching practice ($SD = 0.5$, range from 3.5 to 6 years) from TEDS-Follow-Up ($n = 146$) and taken in parts from TEDS-Instruct and from TEDS-Validate ($n = 47$)—to assure a better discriminability between the groups—constituted the group of early career teachers. According to the Berliner model (1988), they can be classified as being competent at least; some may already have attained proficiency. The third group ($n = 154$) comprised experienced teachers from TEDS-Instruct and TEDS-Validate, who had on average 19.6 years of teaching experience ($SD = 10.4$, range from 6.5 to 41.5 years) and therefore represented participants at the competent, proficient, and expert levels. It can be assumed that the latter group included more proficient and

[2] The master's students had not experienced any extensive teaching practice and internships in their teacher education yet. Whether some students participated in teaching on their own was not asked and could therefore not be considered.

Table 4.1 Characterization of the three expertise groups

Variable	Expertise groups						Test of group differences	
	Master's Students		Early Career Teachers		Experienced Teachers		χ^2 (*df*)/ odds ratio	*p*
Sample size	110		193		154			
Gender (% female)	66.4		58.3		54.5		3.76 (2)	.152
Teaching type (% academic track)	40.0		52.3		45.5		6.72 (2)	.035
	M	*SD*	*M*	*SD*	*M*	*SD*		
Age (in years)	24.57	4.08	31.68	5.50	46.00	9.54	488.36	<.001
Teaching experience (in years)	---	---	4.61	.50	19.64	10.41	397.45	<.001
GPA (Abitur)	2.09	.63	2.11	.61	---	---	21.47	.742
Grade (University)	---	---	1.87	.54	1.95	.61	91.31	<.001
Grade (induction phase)	---	---	1.96	.61	1.98	.67	48.48	.022

Note. GPA—Grade point average; German grades range from 1 (best) to 4 (passed). For the test of group differences in some cases, Fisher's exact test was conducted, since the assumptions for a χ2-test were violated.

expert teachers based on their more extensive teaching practice and reflection. As mentioned, however, experience alone is insufficient to constitute expertise (Caspari-Sadeghi & König, 2018; Palmer et al., 2005; Stigler & Miller, 2018). Rather, experience constitutes an essential factor for expertise. The distinguished groups represent at least three different stages in expertise development, and, therefore, they facilitate an expert–novice comparison. Moreover, the comparison between the two in-service groups may yield valuable insights into the possible impact of long-term teaching experience on expertise development. Table 4.1 presents further characterization of these expertise groups, including grades from university entrance examinations, initial teacher education and the second phase of teacher education (so-called induction phase), the types of schools in which the teachers were contemporarily teaching, and whether the schools were academically oriented with higher achieving students or a more comprehensively oriented track for students of all ability levels.

4.2.3.2 Assessment Instrument

The video-based test instrument used to measure teacher professional noticing limited to secondary education was developed within the TEDS-FU study (Kaiser et al., 2015; Kaiser et al., 2017). It comprises three scripted video-vignettes––Frog King, connected to a German fairy tale, Box, and Solids—of around three and a half minutes in length. The video-vignettes cover a wide range of mathematical topics (e.g., functions, surface, volume calculations, and modeling), different teaching phases, and different school types and showed lessons in the 9th and 10th grades (student age: 14–15 years). During the test, the participants first received some background information about the learning setting, the students, the previous lessons, and the mathematical topic before watching each video-vignette. For instance, the covered mathematical task and its solution were presented. The participants were then permitted to watch the video-vignette simulating real classroom situations once. Directly after watching the video-vignette, participants received open-response and closed (Likert-type) questions, which tested their abilities in all three noticing facets—perception, interpretation, and decision-making—with a mathematics pedagogical or general pedagogical focus. In total, 77 items were included in the tests; seven of these were excluded during scaling owing to their weak psychometric quality that rendered them unsuitable for the samples that were combined specifically for use in this comparative study. Thus, the test instrument used for this study contains 70 items, 36 closed items with Likert scales, and 34 open-response items. The items focused on all three facets of noticing (see Table 4.2), with the lowest number of items focusing on decision-making. The low number of decision-making items is due to their complexity in the construction, as they must be formulated with strong restrictions to be unambiguous. Overall, the testing lasted about 60 minutes in total, with 15–20 minutes per video-vignette.

The test used video-vignettes rather than live observations of the participants in classroom for two reasons. First, this way of data collecting would not have been manageable for the intended sample size, and, second, the use of video-vignettes allows to create situations that are similar to practice, give comparable and standardized test results, and simultaneously present a controllable test environment and a strong connection to performance in classroom (Hughes & Huby, 2001; Piwowar et al., 2018). Moreover, this sort of instrument is partially analagous to situational judgement tests, whose functioning and validity are well examined (Lievens et al., 2008). The research team decided to use scripted videos instead of real teaching footage to assure a density of noteworthy events and provide impressions of a whole teaching lesson in a feasible test duration time.

Table 4.2 Number of items per noticing facet

Noticing facet	Number of items used	Number of rating scale items	Number of open response items
Perception	22	17	5
Interpretation	37	19	18
Decision-making	11	0	11

An expert rating was conducted during test instrument development to determine which answer could be regarded as correct with respect to the rating scales. A coding manual was developed to assess the open-response items and piloted to improve its reliability and validity before it was used in TEDS-FU. For each item, the manual detailed which elements must be present in the response to be coded as correct as well numerous example-answers for correct and incorrect answers as well as border cases gathered in the pilot study. Furthermore, the coding rules left space for minor interpretation, which facilitates a low-inferent coding. Various approaches, including curricular analyses of the mathematical content and comprehensive expert workshops, were employed to ensure the validity of the instrument content and authenticity of the video-vignettes (for details, see Hoth et al., 2016; Kaiser et al., 2015).

The following presents three items and their coding to provide a prototypical insight on the items used and the coding process. In the video-vignette called Solids, the lesson concerns the calculation of the surface and volume of a church tower comprising a cylindrical tower and conical roof.2 The students in the video-vignette calculated the surface and volume, first individually and then in pairs, while the teacher walked through the class helping them. An example for a closed rating scale item concerning this video-vignette asks the participants whether most students participated in the lesson, testing the selective perception of participants regarding classroom events. As given by the expert rating, the correct answer is fully correct, since nearly every student is involved in the lessons. All other options—partially correct, partially incorrect, and not correct at all—were coded as incorrect. Every rating-scale item used these same four response options.

At the end of the video-vignette, the calculation of one student, which contained typical student errors connected to the unreflective usage of the algorithms for surface calculations of solids, is shown. Among other tasks, the study participants had to identify and describe these errors in items concerning the noticing facet of interpretation. Two other open-response items (see Fig. 4.2) challenged them to vary the task set by the teacher in such a way that it has a stronger connection to the real world and at the same time requires modeling skills to be completed. For a coding as correct for the real-world item, participants had to change the task in such a way that it obtained extra-mathematical relevance, like speaking about refurbishing the church tower. For the item concerning modeling, it was necessary to include parts that deal with the transition between reality and mathematics, like a needed simplified mathematical model, before starting to calculate. Both items would correspond to the noticing facet of decision-making, as the participants had to design a suitable task variation to foster certain competencies. Asking for task variation relates the tasks to mathematics pedagogy and school practice. Moreover, mathematics pedagogical content knowledge is required in relation to realistic task setting and mathematical modeling.

A church tower is composed of a cylinder and a cone sitting on it. Both have a radius of 5 m. The cylinder has a height of 25 m, the height of the cone is 12 m. Determine the surface and volume of this composite body.

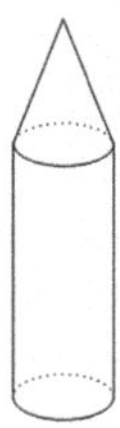

The task used (see left) can be varied in such a way that on the one hand the question becomes more realistic and on the other hand modeling skills are required to complete the task.

Enter a variation of the task in question according to these demands in keywords.

Fig. 4.2 Open-response example item concerning the facet decision making

4.2.3.3 Scaling and Data Analysis

The data analysis comprised the following steps for all samples. First, the open-response items were coded according to the coding manual's rubrics. Two independent raters coded 20% of the test booklets, which means they randomly selected 20% of participants' response to all items, to establish intercoder agreement and ensure reliability. As Table 4.3 indicates, Cohen's kappa values for intercoder agreement were good for all items (Landis & Koch, 1977): between the four studies, the average mean ranged from .80 to .91, and the kappa values ranged from .42 to 1.00. However, few items (four in TEDS-V and seven in

TEDS-V-T) scored an intercoder reliability below 0.61, and at least substantial agreement ($\kappa > 0.61$) was achieved for nearly every item. The displayed kappa values for TEDS-V and TEDS-V-T were adjusted. For these studies, one (TEDS-V) and five values (TEDS-V-T) were excluded, because these items showed poor intercoder reliability, particularly due to low incidence of correct answers. These items were again discussed by the raters and then coded by consensus. The final calculation of Cohen's kappa, for which descriptive results are provided in Table 4.3, excluded these items.

Table 4.3 Intercoder reliability

Study	κ_{mean}	κ_{min}	κ_{max}
TEDS-FU	.86	.76	1.000
TEDS-I	.91	.70	1.000
TEDS-V	.85	.42	1.000
TEDS-V-T	.80	.47	1.000

Note. κ = Cohens Kappa. Inter-coder agreement was investigated for all 34 open-response items and each study. Values above 0.4/0.6/0.8 indicate moderate/substantial/almost perfect agreement (Landis & Koch, 1977). Values for TEDS-V and TEDS-V-T were adjusted.

The data were scaled using an item response theory (IRT) approach with the software ConQuest Version 4.5.2 (Wu et al., 1997). First, a missing response was counted as not administered to estimate item parameters only on the basis of valid answers. Second, missing responses were recoded as wrong answers for ability estimation.[3] To validate the hypothesis that three scales for teacher noticing facets were required, a corresponding three-dimensional IRT model was calculated and compared with a one-dimensional IRT model that would not differentiate into noticing facets. To facilitate reading and interpretation, ability estimates obtained through scaling analyses were linearly transformed to a mean of 50 and a standard deviation of 10. Further analysis was conducted using chi-square tests, Fisher's exact test, and Scheffé post hoc tests to investigate significant differences between expertise groups. Measurement-error-free correlation analyses were carried out within the IRT model to examine associations between the facets.

[3] Missing amount per item ranged from 2 to 16% with a mean of 5%. Open-response items showed significantly ($p < .001$) more missings than rating-scale items. Participants had 5% missings on average with a range from 0 to 48%. Master's students displayed 10% missings on average, and early career teachers and experienced teachers both displayed 3%. Hence, master's students produced significantly ($p < .001$) more missings than teachers.

4.2.4 Results

We will now present the results in two sections, where each section addresses one of our research questions.

4.2.4.1 Empirical Separation of the Facets of Noticing and its Association

A one-dimensional Rasch model was compared with a three-dimensional model to investigate the hypothesis of three separate scales. Two established indices of fit—the ratio of chi-square to degrees of freedom and the Akaike information criterion (AIC)—were then calculated (Table 4.4). They revealed a significantly better fit of the three-dimensional model for the data and confirmed the validity of separating the noticing construct into three facets.[4]

Table 4.4 Fit indices for both model approaches

Model	χ^2	χ^2/df	AIC	Δ AIC
one-dimensional	5,047.27	73.15	39,247	
three-dimensional	4,698.45	70.13	39,192	−55

Note. AIC difference is relative to the one-dimensional model. The difference of the models' deviances is significant with $p < 0.001$. AIC = Akaike information criterion

As Table 4.5 indicates, the three-dimensional model estimates the three facets with at least acceptable reliability. The item-total correlation of this model varied between .12 and .50 with an acceptable mean of .31. The weighted mean square, which is part of the fit statistics and has an expected value of 1.0, ranged from .90 to 1.10 with a mean of 1.00. Values between .50 and 1.50 indicate a productive fit (Linacre, 2002).

The variation of item difficulty in relation to the individuals' abilities (Fig. 4.3) showed good estimation of teacher noticing abilities, although difficult items were somewhat underrepresented.

The data illustrate that the three facets of noticing conceptualized as perception, interpretation, and decision-making can be empirically separated and measured with at least acceptable reliability for the different reliability measures (except WLE for decision-making) using separate scales (Table 4.5).

[4] For χ^2, χ^2/df, and AIC, the better model is shown by a lower value. To deepen the model comparison, additional calculations using structural equation modeling were computed and are displayed in the Appendix.

Table 4.5 Scale reliability for all three facets

Reliability measure	Reliability		
	Perception	Interpretation	Decision-making
EAP	0.704	0.800	0.621
WLE	0.663	0.796	0.489

Note. EAP (expected as posteriori) and WLE (weighted likelihood estimates) are reliability measures provided by the software ConQuest (Wu et al., 1997) and can be interpreted similar to Cronbach's Alpha.

Overall, the empirical results indicate substantial associations between the three facets of teacher noticing. The latent correlations between perception and interpretation ($r = 0.814$) and those between interpretation and decision-making ($r = 0.815$) are relatively strong and can be evaluated as a large positive effect or association (Döring & Bortz, 2016). A medium correlation between perception and decision-making ($r = 0.462$) was also identified (see Table 4.6). Thus, all three measured facets can be understood as associated, which empirically supports statements regarding the interrelated character of the noticing construct in the noticing discussion (e.g., Sherin et al., 2011b). The high correlation between perception and interpretation emphasizes the strong connection between these two noticing facets as perception is a cognitive, sense-based process, which can only be done in a limited way without interpretation. Although efforts were made to construct items that measured only one facet, namely perception or interpretation, it is hardly possible to adequately differentiate between them. In addition, the strong correlation between interpretation and decision-making can be interpreted as further justification for the inclusion of an action-related facet in the concept of teacher noticing and the theoretical intertwining of both facets.

Overall, this means that the three facets are indeed empirically separable, although the correlations between them are relatively high, particularly between perception and interpretation as well as between interpretation and decision-making. Whether group dependency or group-specific patterns characterize the correlations was not investigated in the analyses of this study and needs to be subject of further studies.

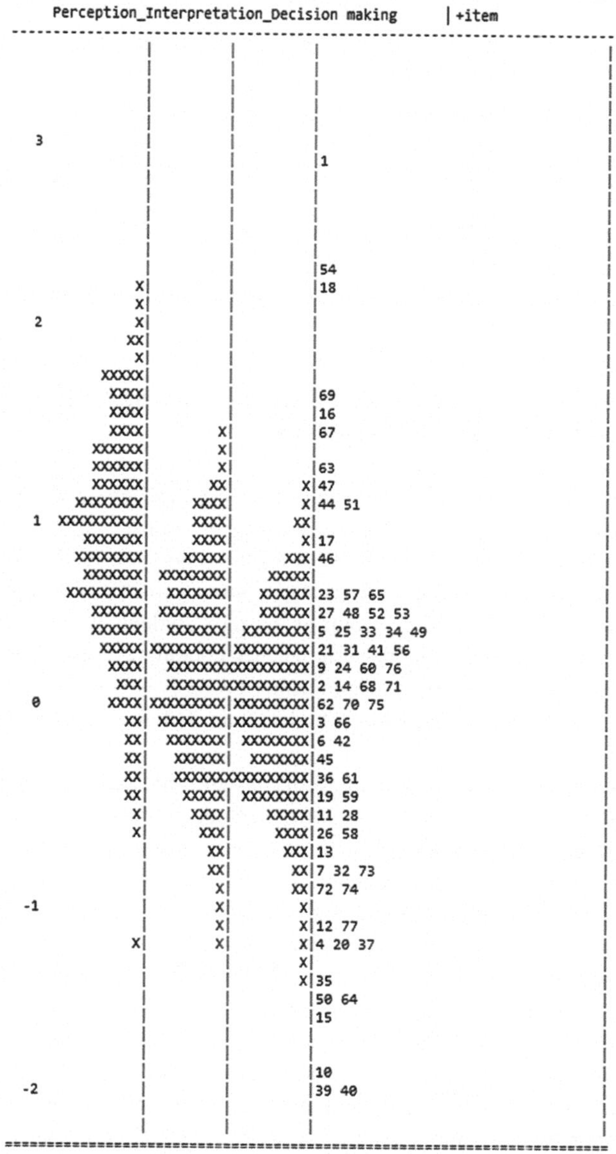

Note. Figure is presented here as given by ConQuest (Wu et al., 1997). Personal ability values were transformed to $M = 50$, $SD = 10$ afterwards. One X represent 3.7 cases. Numbers 1 to 77 represent the respective items.

Fig. 4.3 Variation of the item difficulty in relation to participants' ability for all three scales

Table 4.6 Variance (on the diagonal), covariance (above the diagonal) and correlation (below the diagonal) for all three facets

Facet	Variance/ Covariance/Correlation		
	Perception	Interpretation	Decision-making
Perception	.455	.329	.159
Interpretation	.814	.359	.248
Decision-making	.462	.815	.258

Note. Correlations were effectively corrected for any attenuation caused by measurement error.

4.2.4.2 Development of Teacher Noticing Across Different Teacher Groups of Expertise

To address the second research question on the differences in teacher noticing and its facets between groups of pre- and in-service teachers, a cross-sectional comparison was conducted for each facet across the groups.

Significant patterns can be identified for all three noticing facets (Table 4.7). For all facets, significant mean differences in scores emerged as indicated by the Scheffé post hoc test from the group of master's students to the early career teachers, indicating a possible increase in teacher noticing average scores. However, the mean score of the experienced teacher group is slightly lower than that of the early career teachers but clearly higher than that of the master's student mean.

These results suggest that noticing skills progressed significantly from the group of master's students to early career teachers, demonstrating the importance of teaching practice for the development of expertise in noticing. Furthermore, stagnation at least or even small regression is noted in the comparison of early career teacher noticing and experienced teacher noticing. After nearly five years of professional teaching practice, early career teachers seem to perform better than experienced teachers with 20 + years of professional experience. However, as shown by the Scheffé post hoc test, these results only were significant for the decision-making facet. To investigate the connection between teaching experience and teacher noticing even further, the team conducted linear regression analyses with teaching experience as independent and each of the three noticing facets as dependent variables using the software Mplus Version 8.6 (Muthén & Muthén, 2021). Given that the master's students had no regular extensive teaching practice yet, only the data of the in-service teachers, that is, the group of early career and experienced teachers, were included. As shown in Table 4.8, the regression

Table 4.7 Mean scores by expertise group and facet

Variable	Expertise groups						Test of group differences		
	MS ($n = 110$)		ECT ($n = 193$)		ET ($n = 154$)		odds ratio	p	Scheffé post-hoc test[b]
	Mean	SD	Mean	SD	Mean	SD			
Perception[a]	45.57	11.08	51.84	9.19	50.07	8.90	101.69	<.001	ET > MS, ECT > MS
Interpretation[a]	43.58	12.01	52.62	8.44	50.99	8.45	127.90	<.001	ET > MS, ECT > MS
Decision-making[a]	44.04	11.21	53.11	8.36	50.27	9.29	94.46	<.001	ET > MS, ECT > MS, ECT > ET

Note. MS = master's students, ECT = early career teachers, ET = experienced teachers, SD = standard deviation. For the test of group differences, a Fisher's exact test was conducted, since the assumptions for a χ2-test were violated.
[a] $M = 50$ and $SD = 10$.
[b] All given differences in the Scheffé post-hoc test are statistically significant ($p < .05$). More statistical information to the Scheffé post-hoc test is given in the appendix

reveals a small, statistically significant ($p < .01$) decline with increasing teaching experience for all three facets ($\beta_{perception} = -.16$, $\beta_{interpretation} = -.22$, $\beta_{decision-making} = -.21$)[5]. The group of teachers with low experience was somewhat overrepresented, as participants of the TEDS-FU study with low experience represented a distinct part of the sample, which violated the assumptions of the regression. A regression excluding the group of early career teachers revealed similar significant findings.

[5] Scatter plots are displayed in the Appendix.

Table 4.8 Linear regression of perception, interpretation and decision-making on teaching experience

Variable	Perception			Interpretation			Decision-making		
	b	*β*	*p*	*b*	*β*	*p*	*b*	*β*	*p*
Teaching experience	–.14	–.16	<.01	–.18	–.22	<.001	–.18	–.21	<.001
Intercept	51.06	5.62	<.001	51.90	6.13	<.001	51.85	5.85	<.001
R^2		.025			.048			.044	

4.2.5 Summary, Discussion, and Conclusions

The present study aimed to investigate two questions—whether the concept of noticing can be assessed analytically using three facets and whether there are mean differences in teacher noticing among different groups of mathematics teachers with different amounts of professional teaching experience.

Concerning the first research question, the study findings highlight that noticing can be measured analytically and it is possible to measure the three facets of noticing—perception, interpretation, and decision-making—separately through reliable scales based on the rigorous standards of quantitative educational research. However, the results highlight that the three facets of teacher noticing are highly intercorrelated, as was suggested theoretically in the noticing discussion. In particular, the facets of perception and interpretation are strongly connected, which is not unexpected as it is difficult to perceive without interpretation. Furthermore, the strong association between interpretation and decision-making was anticipated as noticing has been described as a knowledge-based construct in the literature. Decision-making has been conceptualized as developing proposals for further continuations of the lesson or alternative ideas, which should be based on a sound interpretation of the noticed situation. The construct of noticing can therefore be divided into three different facets that are strongly connected as theoretically expected and explainable. Furthermore, a new facet, decision-oriented, emerged, which has the potential to enrich the theoretical construct of noticing and offers an empirical underpinning for our theoretical framework.

The second research question focused on differences in noticing among different pre- and in-service teacher groups with different professional teaching experience and likely also different levels of expertise. Comparisons between the different groups, ranging from master's students to practicing teachers with different degrees of teaching experience, allowed us to assume a progression, which can be interpreted as evidence for the relevance of professional teaching experience. However, the data only show that master's students were significantly outperformed by the two other groups, whereas experienced teachers did not outperform early career teachers. This pattern was identified for all three facets, indicating an underlying pattern.

One reason for the slight decrease in noticing skills from the cross-sectional evaluated groups of early career to experienced teachers may be a "falling-off of skill" (Forde & McMahon, 2019, p. 142) among experienced teachers, which might be explained by the fact that teachers with over 20 years of experience are less likely to participate in professional development (Pedder et al., 2008). Additionally, the expectation of linear growth could be challenged, given that, as mentioned by Berliner (1988), only some teachers attain proficiency and even fewer will attain expert status. The slight regression might indicate that teaching experience in years may contribute to expertise along with other factors, such as working in teacher education or the promotion of strong students' achievements (Palmer et al., 2005), which may vary across both practicing teacher groups. These findings may also substantiate that, as Patterson (2019, p. 4) hypothesizes, teacher personal professional development "may not be linear or truly cyclic but may have stops, starts, digressions, and regressions at various times." However, this may also provide evidence for a general, non-individual-linked pattern of development and raises questions about the long-term effects of teacher education (Liu & Phelps, 2020).

This result is, on the one hand, unexpected, referring to the findings of Jacobs et al. (2010) and Yang et al. (2021b) who interpret their findings as a monotonic or nearly linear growth in noticing between the various expertise groups covering similar teacher groups. The difference in the results reported by Jacobs et al. (2010) may be attributed to the fact that in this study, teacher educators were included in the final stage. The study by Yang et al. (2021b) is embedded in a cultural background—Chinese teachers coming from an Eastern educational paradigm—that differs from this study of German teachers reflecting a Western educational paradigm. The strong focus on content-related and pedagogical aspects has led to the description of newly graduated teachers in China as "semi-finished products" (Paine et al., 2003, p. 216), who will need to learn and develop their educational skills when they enter schools as recent graduates. New graduates are accompanied by mentors, whose role is to impart lessons to the new teachers and work to improve individual lessons or teaching units (Lu et al., 2020). By contrast, the seminar-based second phase of teacher education in Germany (the so-called induction phase) introduces university graduates to teaching on a theoretical basis, observing joint lessons, and theory-oriented phases and topics and aims to induct the candidates into teaching practice from a reflective perspective (Landesinstitut für Lehrerbildung und Schulentwicklung, 2015, 2019). Overall, these different ways of inducting new teachers into teaching may lead to different growth patterns across the three expertise groups.

The findings of this study also align with other empirical studies of teacher knowledge. Evidence suggests a strong increase in GPK during initial teacher education (e.g., König, 2013) and a flattening of the growth after entering the second phase of teacher education (Germany) or the teaching profession as early career teachers (Austria) (König et al., under review). Furthermore, the results of a study by Kleickmann et al. (2013) show a significant increase in MCK and MPCK for teachers from the beginning of initial teacher education to the end of the second phase of teacher education (induction phase). However, their study results describe stagnation or decrease tendencies for MCK and also only slight progression or stagnation for MPCK in both groups when early career teachers and experienced teachers are compared. Both studies report longitudinally interpreted cross-sectional data. Similar results are presented by Blömeke et al. (2014), referring to longitudinal data from the TEDS-M and TEDS-FU Study.

To summarize, the global stagnation or small decrease effects between both groups of practicing teachers indicate that teaching practice is not the only decisive factor in the development of teacher noticing, and the mere extension of teaching experience does not automatically yield higher achievements in noticing. The results provided clear evidence for the necessity of professional development, which would allow more senior teachers to learn more about recent didactical approaches (e.g., Lipowsky, 2019).

Finally, the study limitations must be acknowledged. The analyses presented here were based on convenience samples and cross-sectional data, and caution is urged regarding the development of teacher noticing. Furthermore, the homogeneous TEDS-FU sample, which constitutes most of the early career teachers, may have advantaged this group. One reason for the regression between early career and experienced teachers may be that the sample of early career teachers from TEDS-FU was relatively homogenous and possibly cognitively more selective than the other groups (Blömeke et al., 2014).

Furthermore, apart from the linear regressions, which regarded more variance, only three different expertise groups were compared, making it difficult to establish differences by the levels of expertise outlined by Berliner (1988). Overall, further studies with more differentiated groups are needed to better understand the development of expertise in teacher noticing. In particular, the comparison of groups with different cultural backgrounds is necessary, as the results suggest cultural influences on the development of teachers professional noticing. These differences must be addressed if Eastern and Western cultures wish to cooperate and learn from one another as has occurred in the past (Kaiser & Blömeke, 2013).

4.2.6 Appendix

4.2.6.1 Examination of the Noticing Conceptualization with Structural Equation Modeling (in Addition to *Table 4.4*)

To deepen the model comparison, additional calculations using structural equation modeling were computed. We compared a one-dimensional model that used all 70 items as indicators for the latent trait teacher noticing with a three-dimensional model, where the respective items were indicators for the three latent traits perception, interpretation and decision-making. The calculation was implemented with Mplus Version 8.6 (Muthén & Muthén, 2021) using the robust maximum likelihood estimators MLR (Table 4.9). It states the better fit of the three-dimensional model for all common information criterions.

Since the chi-square test could not be computed with MLR estimators due to a too large frequency table, weighted least square mean and variance adjusted estimators (WLSMV) were also applied to conduct a chi-square test for differences and calculate the root mean square error of approximation (RMSEA) and other fit indices to investigate the model fit (Table 4.10). Both models fit the data, but the chi-square test for model differences shows clear advantages of the three-dimensional model. Comparative fit index (CFI) shows only poor fit, which was expected since it punishes complex models with large number of indicators relative to sample size such as were used in our IRT approach. Besides, as we examined our data in a confirmatory context, for evaluating the models, we prefer RMSEA following recommendations by Rigdon (1996).

Table 4.9 Fit indices for both model approaches using MLR estimators

Fit indices	one-dimensional model	three-dimensional model	Δ
AIC	36,531	36,470	–61
BIC	37,133	37,085	–48
Adjusted BIC	36,670	36,612	–58

Note. Differences are relative to the one-dimensional model. For AIC, BIC and Adjusted BIC the better model is shown by a lower value. AIC = Akaike information criterion, BIC = Bayesian information criterion. Adjusted BIC were adjusted by sample size.

Table 4.10 Fit indices for both model approaches using WLSMV estimators

Fit indices	one-dimensional model	three-dimensional model	Δ
$\chi 2$	3,269.868	3,234.898	–45.483[a]
RMSEA	.029 [.027, .032]	.029 [.026, .031]	---
CFI	.646	.658	.012

Note. Differences are relative to the one-dimensional model. $\chi 2$ = chi-square test of model fit, RMSEA = root mean square error of approximation, CFI = comparative fit index. For RMSEA the 90-percent-conficence-intervall is given in brackets. Cut-off for RMSEA is .05. a Chi-square test for difference testing used the diff-test option. Therefore, the difference does not result from both individual values.

4.2.6.2 Linear Regression for Perception, Interpretation, and Decision-Making by Teaching Experience (in Addition to *Tables 4.7 and 4.8*)

For all three noticing facets, linear regressions with the noticing facet as dependent variable and teaching experience in years as independent variable were conducted (see Figures 4.4, 4.5, and 4.6). Since the master's students had no regular extensive teaching practice yet, only the data of the in-service teachers, i.e., the group of early career and experienced teachers, were included.

4.2.6.3 Additional Data for the Scheffé post hoc Test (in Addition to *Table 4.7*)

We present all statistical data for the Scheffé post hoc test in Table 4.11.

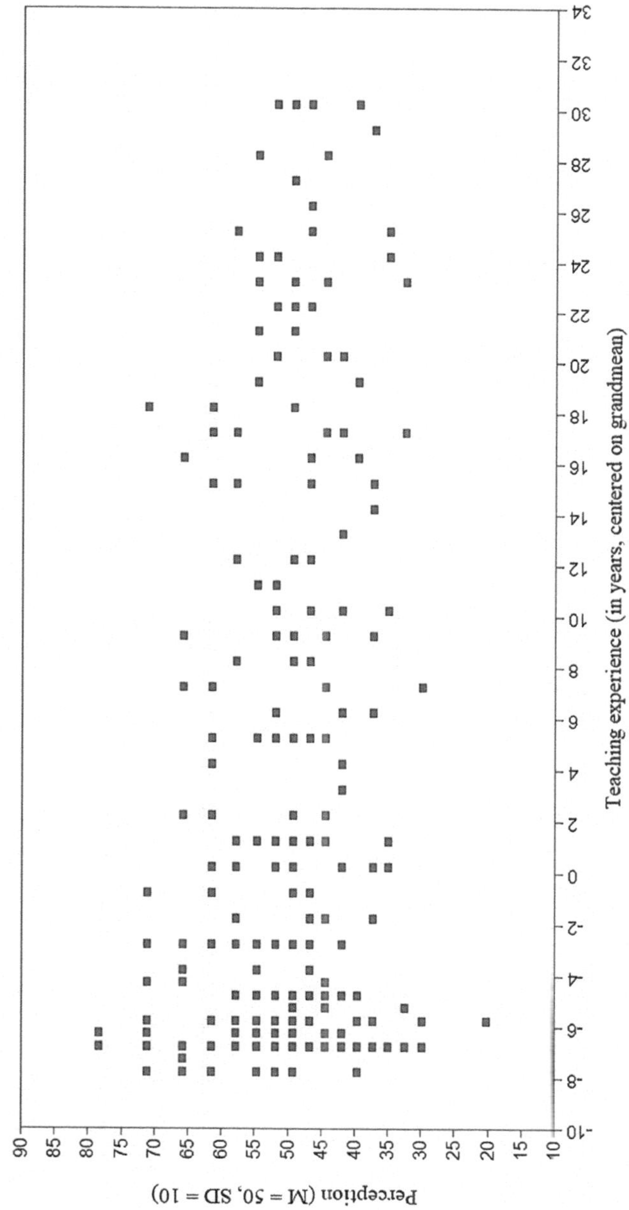

Note. Teaching experience is centered on the mean ($M = 11.28$), which, therefore, functions as zero in years of teaching experience.

Fig. 4.4 Scatter plot of perception (Y) vs. teaching experience (X)

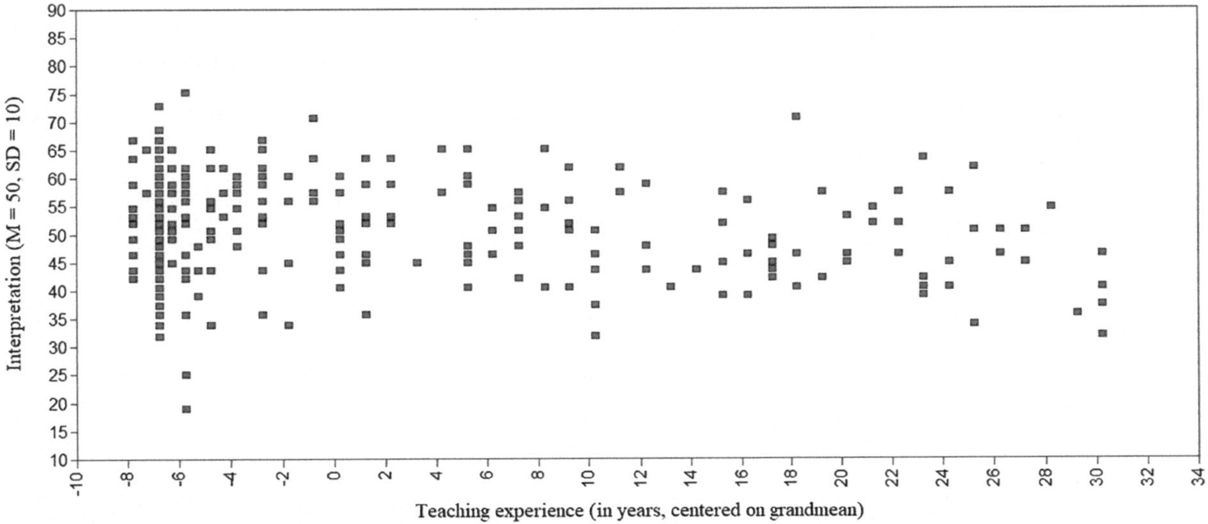

Note. Teaching experience is centered on the mean ($M = 11.28$) which, therefore, functions as zero in years of teaching experience.

Fig. 4.5 Scatter plot of interpretation (Y) vs. teaching experience (X)

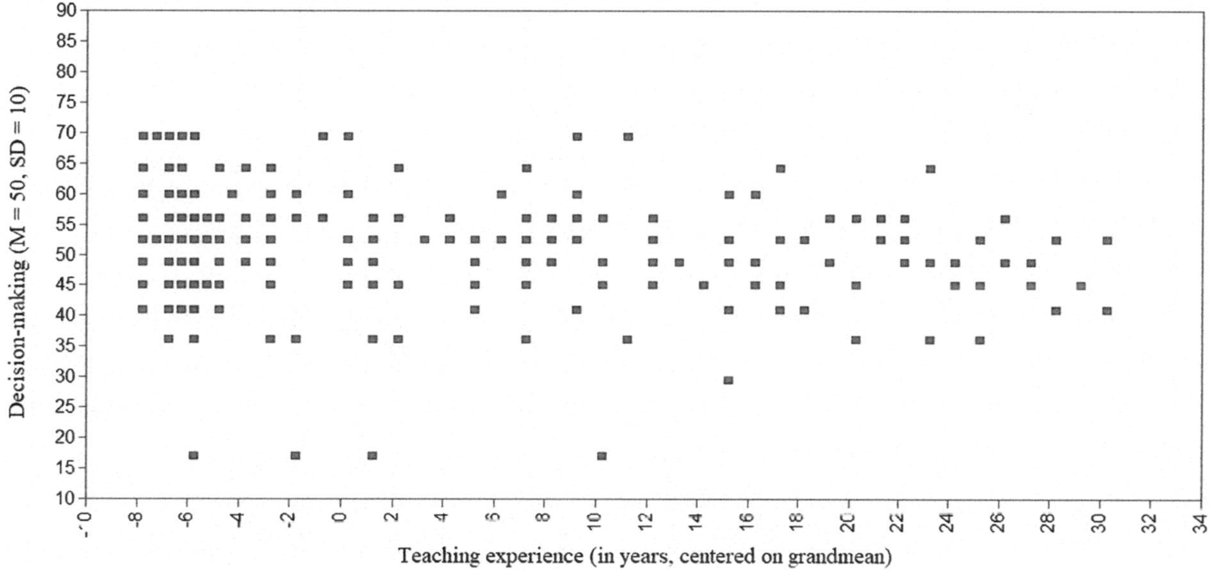

Note. Teaching experience is centered on the mean ($M = 11.28$) which, therefore, functions as zero in years of teaching experience.

Fig. 4.6 Scatter plot of decision-making (Y) vs. teaching experience (X)

Table 4.11 Additional data for the Scheffé post-hoc test

Group a	Group b	Mean difference (a-b)	*SE*	*p*	95% CI
Perception					
MS	ECT	–6.27	1.15	<.001	[–9.08, –3–46]
MS	ET	–4.5	1.20	.001	[–7.44, –1.56]
ECT	ET	1.77	1.04	.233	[–.77, –4.31]
Interpretation					
MS	ECT	–9.03	1.13	<.001	[–11.80, –6.27]
MS	ET	–7.41	1.18	<.001	[–10.30, –4.52]
ECT	ET	1.62	1.02	.282	[–.88, 4.12]
Decision-making					
MS	ECT	–9.07	1.13	<.001	[–11.84, –6.31]
MS	ET	–6.24	1.18	<.001	[–9.13, –3.35]
ECT	ET	2.83	1.02	.021	[.33, 5.33]

Note. MS = master's students, ECT = early career teachers, ET = experienced teachers, SE = standard error, CI = confidence interval.

Literature

Rigdon, E. E. (1996). CFI versus RMSEA: A comparison of two fit indexes for structural equation modeling. Structural Equation Modeling: A Multidisciplinary Journal, 3(4), 369–379.

Funding:

Open Access funding enabled and organized by Projekt DEAL.

TEDS-Follow-Up: German Research Foundation (grant number: KA 797/18–1)

TEDS-Validate: Federal Ministry of Education and Research (grant number: 01PK15006A/B)

TEDS-Validate-Transfer: Federal Ministry of Education and Research (grant number: 01PK19006A/B)

Declarations

Conflict of interest: The authors declare no competing interests.

References

Amador, J. M., Bragelman, J., & Superfine, A. C. (2021). Prospective teachers' noticing: A literature review of methodological approaches to support and analyze noticing. *Teaching and Teacher Education*, 99, 103256. https://doi.org/10.1016/j.tate.2020.103256

Berliner, D. C. (1988). *The development of expertise in pedagogy*. American Association of Colleges for Teacher Education.

Berliner, D. C. (1994). Expertise: The wonders of exemplary performance. In J. N. Mangieri & C. Collins (Eds.), *Creating powerful thinking in teachers and students* (pp. 141–186). Holt.

Berliner, D. C. (2001). Learning about and learning from expert teachers. *International Journal of Educational Research*, 35(5), 463–482. https://doi.org/10.1016/S0883-0355(02)00004-6

Berliner, D. C. (2004). Expert teachers: Their characteristics, development and accomplishments. In R. Batllori i Obiols, A. E. Gomez Martinez, M. Oller i Freixa, & J. Pages i Blanch (Hrsg.), *De la teoria… a l'aula: Formacio del professorat ensenyament de las ciències socials* (pp. 13–28). Departament de Didàctica de la Llengua de la Literatura I de les Ciències Socials, Universitat Autònoma de Barcelona.

Blömeke, S., Busse, A., Kaiser, G., König, J., & Suhl, U. (2016). The relation between content-specific and general teacher knowledge and skills. *Teaching and Teacher Education*, 56, 35–46.

Blömeke, S., Gustafsson, J. E., & Shavelson, R. J. (2015). Beyond dichotomies: Competence viewed as a continuum. *Zeitschrift für Psychologie*, 223(1), 3–13.

Blömeke, S., König, J., Busse, A., Suhl, U., Benthien, J., Döhrmann, M., & Kaiser, G. (2014). From teacher education into profession: Subject-related knowledge as prerequisite for perception, analysis and reaction in teaching [Von der Lehrerausbildung in den Beruf: Fachbezogenes Wissen als Voraussetzung für die Wahrnehmung, Analyse und Reaktion im Unterricht]. *Zeitschrift für Erziehungswissenschaft*, 17, 509–542.

Blömeke, S., Suhl, U., Kaiser, G., & Döhrmann, M. (2012). Family background, entry selectivity and opportunities to learn: What matters in primary teacher education? An international comparison of fifteen countries. *Teaching and Teacher Education*, 28, 44–55.

Carter, K., Cushing, K., Sabers, D., Stein, P., & Berliner, D. (1988). Expert-novice differences in perceiving and processing visual classroom information. *Journal of Teacher Education*, 39(3), 25–31. https://doi.org/10.1177/002248718803900306

Caspari-Sadeghi, S., & König, J. (2018). On the adequacy of expert teachers: From practical convenience to psychological reality. *International Journal of Higher Education*, 7(5), 1–19. https://doi.org/10.5430/ijhe.v7n5p1

Dindyal, J., Schack, E. O., Choy, B. H., & Sherin, M. G. (2021). Exploring the terrains of mathematics teacher noticing. *ZDM Mathematics Education*, 53(1), 1–16. https://doi.org/10.1007/s11858-021-01249-y

Dominguez, H. (2019). Theorizing reciprocal noticing with non-dominant students in mathematics. *Educational Studies in Mathematics*, 102(1), 75–89.

Döring, N., & Bortz, J. (2016*). Forschungsmethoden und Evaluation in den Sozial- und Humanwissenschaften* [Research methods and evaluation in the social and human sciences] (5th ed.). Springer. https://doi.org/10.1007/978-3-642-41089-5

Dreher, A., & Kuntze, S. (2015). Teachers' professional knowledge and noticing: The case of multiple representations in the mathematics classroom. *Educational Studies in Mathematics*, 88(1), 89–114. https:// doi.org/https://doi.org/10.1007/s10649-014-9577-8

Erickson, F. (2011). On noticing teacher noticing. In M. G. Sherin, V. R. Jacobs, & R. A. Philipp (Eds.), *Studies in mathematical thinking and learning. Mathematics teacher noticing* (pp. 17–34). Routledge.

Ericsson, K. A., Krampe, R. T., & Tesch-Römer, C. (1993). The role of deliberate practice in the acquisition of expert performance. *Psychological Review*, 100, 363–406.

Forde, C., & McMahon, M. (2019). *Teacher quality, professional learning and policy: Recognising, rewarding and developing teacher expertise.* https://doi.org/10.1057/978-1-137-53654-9

Friesen, M., & Kuntze, S. (2021). How context specific is teachers' analysis of how representations are dealt with in classroom situations? Approaching context-aware measure for teacher noticing. *ZDM Mathematics Education*, 53(1), 181–193. https://doi.org/10.1007/s1185.8-020-01204-3

Gold, B., & Holodynski, M. (2017). Using digital video to measure the professional vision of elementary classroom management: Test validation and methodological challenges. *Computers & Education*, 107, 13–30. https://doi.org/10.1016/j.compedu.2016.12.012

Goodwin, C. (1994). Professional vision. *American Anthropologist*, 96(3), 606–633.

Hoth, J., Schwarz, B., Kaiser, G., Busse, A., König, J., & Blömeke, S. (2016). Uncovering predictors of disagreement: Ensuring the quality of expert ratings. *ZDM Mathematics Education*, 48(1–2), 83–95. https://doi.org/10.1007/s11858-016-0758-z

Huang, R., & Li, Y. (2012). What matters most: A comparison of expert and novice teachers' noticing of mathematics classroom events. *School Science and Mathematics*, 112(7), 420–432. https://doi.org/10.1111/j.1949-8594.2012.00161.x

Hughes, R., & Huby, M. (2001). The application of vignettes in social and nursing research. *Journal of Advanced Nursing*, 37(4), 382–386. https://doi.org/10.1046/j.1365-2348.200202100.x

Ivars, P., Fernández, C., & Llinares, S. (2020). A learning trajectory as a scaffold for pre-service teachers' noticing of students' mathematical understanding. *International Journal of Science and Mathematics Education*, 18, 529–548. https://doi.org/10.1007/s10763-019-09973-4

Jacobs, V. R., Lamb, L. L. C., & Philipp, R. A. (2010). Professional noticing of children's mathematical thinking. *Journal for Research in Mathematics Education*, 41(2), 169–202 www.jstor.org/stable/20720 130

Kaiser, G., & Blömeke, S. (2013). Learning from the eastern and the western debate: The case of mathematics teacher education. *ZDM Mathematics Education*, 45(1), 7–19.

Kaiser, G., Blömeke, S., König, J., Busse, A., Döhrmann, M., & Hoth, J. (2017). Professional competencies of (prospective) mathematics teachers—Cognitive versus situated approaches. *Educational Studies in Mathematics*, 94(2), 1–22. https://doi.org/10.1007/s10649-016-9713-8

Kaiser, G., Busse, A., Hoth, J., König, J., & Blömeke, S. (2015). About the complexities of video-based assessments: Theoretical and methodological approaches to overcoming shortcomings of research on teachers' competence. *International Journal of Science and Mathematics Education*, 13(2), 369–387. https://doi.org/10.1007/s10763-015-9616-7

Kaiser, G. & König, J. (2019). Competence measurement in (mathematics) teacher education and beyond: Implications for policy. *Higher Education Policy*, 32, 597–615. https://doi.org/10.1057/s41307-019-00139-z

Keppens, K., Consuegra, E., Goossens, M., de Maeyer, S., & Vanderlinde, R. (2019). Measuring pre-service teachers' professional vision of inclusive classrooms: A video-based comparative judgement instrument. *Teaching and Teacher Education*, 78, 1–14. https://doi.org/10.1016/j.tate.2018.10.007

Kleickmann, T., Richter, D., Kunter, M., Elsner, J., Besser, M., Krauss, S. & Baumert, J. (2013). Teachers' content knowledge and pedagogical content knowledge. *Journal of Teacher Education*, 64(1), 90–106. https://doi.org/10.1177/0022487112460398

König, J. (2013). First comes the theory, then the practice? On the acquisition of general pedagogical knowledge during initial teacher education. *International Journal of Science and Mathematics Education*, 11(4), 999–1028. https://doi.org/10.1007/s10763-013-9420-1

König, J. (2015). Measuring classroom management expertise (CME) of teachers: A video-based assessment approach and statistical results. *Cogent Education*, 2(1), 991178.

König, J., Blömeke, S., Jentsch, A., Schlesinger, L., Felske, C., Musekamp, F., & Kaiser, G. (2021). The links between pedagogical competence, instructional quality, and mathematics achievement in the lower secondary classroom. *Educational Studies in Mathematics*, 107, 189–212.

König, J., Ligtvoet, R., Klemenz, S., & Rothland, M. (under review). Discontinued knowledge growth: On the development of teachers' general pedagogical knowledge at the transition from higher education into teaching practice.

König, J., Santagata, R., Scheiner, T., Adleff, A.-K., Yang, X., & Kaiser, G., (under review). Teacher noticing: A systematic literature review on conceptualizations, research designs, and findings on learning to notice.

Lachner, A., Jarodzka, H., & Nückles, M. (2016). What makes an expert teacher? Investigating teachers' professional vision and discourse abilities. *Instructional Science*, 44(3), 197–203. https://doi.org/10.1007/s11251-016-9376-y

Landesinstitut für Lehrerbildung und Schulentwicklung. (2015). *Ausbildungscurriculum Mathematik* [Curriculum for education in mathematics]. https://li.hamburg.de/contentblob/5887082/698b68ec837e795ca8b5daa4364a7b50/data/pdf-ac-mathe.pdf

Landesinstitut für Lehrerbildung und Schulentwicklung. (2019). *Referenzrahmen Lehrkräfteausbildung und Anpassungsqualifizierung* [Framework for teacher education and qualification]. https://li.hamburg.de/contentblob/12779726/f628678e4a1c59.b9614951ddc71d21c6/data/pdf-referenzrahmen-19-08.pdf

Landis, J. R., & Koch, G. G. (1977). The measurement of observer agreement for categorical data. *Biometrics*, 33(1), 159. https://doi.org/10.2307/2529310

Lievens, F., Peeters, H., & Schollaert, E. (2008). Situational judgement tests: A review of recent research. *Personnel Review*, 37 (4), 426–441. https://doi.org/10.1108/00483480810877598

Linacre, J. M. (2002). What do infit and outfit, mean-square and standardized mean? *Rasch Measurement Transactions*, 16(2), 878.

Lipowsky, F. (2019). Wie kommen Befunde der Wissenschaft in die Klassenzimmer? – Impulse der Fortbildungsforschung [How do findings from research get into classroom?—Impulses from professional development research]. In C. Donie, F. Foerster, M. Obermayr, A. Deckwerth, G. Kammermeyer, G. Lenske, M. Leuchter, & A. Wildemann (Eds.), *Grundschulpädagogik zwischen Wissenschaft und Transfer* (pp. 144–161). VS Verlag.

Liu, S., & Phelps, G. (2020). Does teacher learning last? Understanding how much teachers retain their knowledge after professional development. *Journal of Teacher Education*, 71(5), 537–550. https://doi.org/10.1177/0022487119886290

Louie, N. L. (2018). Culture and ideology in mathematics teacher noticing. *Educational Studies in Mathematics*, 97(1), 55–69.

Louie, N. L., Adiredja, A. P., & Jessup, N. (2021). Teacher noticing from a sociopolitical perspective: The FAIR framework for anti-deficit noticing. *ZDM Mathematics Education*, 53(1), 95–107. https://doi.org/10.1007/s11858-021-01229-2

Lu, X., Kaiser, G., & Leung, F. (2020). Mentoring early career mathematics teachers from the mentees' perspective—A case study from China. *International Journal for Science and Mathematics Education*, 18, 1355–1374.

Mason, J. (2002). *Researching your own practice: The discipline of noticing*. Routledge.

Mason, J. (2011). Noticing: roots and branches. In M. G. Sherin, V. R. Jacobs, & R. A. Philipp (Eds.), *Studies in mathematical thinking and learning. Mathematics teacher noticing* (pp. 35–50). Routledge.

Mason, J. (2015). Responding in-the-moment: Learning to prepare for the unexpected. *Research in Mathematics Education*, 17(2), 110–127. https://doi.org/10.1080/14794802.2015.1031272

Meschede, N., Fiebranz, A., Möller, K., & Steffensky, M. (2017). Teachers' professional vision, pedagogical content knowledge and beliefs: On its relation and differences between pre-service and in-service teachers. *Teaching and Teacher Education*, 66, 158–170. https://doi.org/10.1016/j.tate.2017.04.010

Muthén, L. K., & Muthén, B. O. (2021). *Mplus 8* (Version 8.6) [computer software]. Muthén & Muthén http://statmodel.com

Paine, L. W., Fang, Y., & Wilson, S. (2003). Entering a culture of teaching. In E. Britton, L. Paine, D. Pimm, & S. Raizen (Eds.), *Comprehensive teacner induc:ion: Systems for early career learning* (pp. 20–82). Kluwer.

Palmer, D. J., Stough, L. M., Burdenski, T. K., & JR., & Gonzales, M. (2005). Identifying teacher expertise: An examination of researchers' decision making. *Educational Psychologist*, 40(1), 13–25. https://doi.org/10.1207/S15326985EP4001_2

Patterson, C. (2019). *Enacted personal professional learning: Re-thinking teacher expertise with story-telling and problematics*. Springer. https://doi.org/10.1007/978-981-13-6007-7

Pedder, D., Storey, A., Opfer, V. D., & Wolfenden, F. (2008). *Schools and continuing professional development (CPD) in England State of the Nation research project*. Cambridge University and The Open University.

Piwowar, V., Barth, V. L., Ophardt, D., & Thiel, F. (2018). Evidence-based scripted videos on handling student misbehavior: The development and evaluation of video cases for teacher education. *Professional Development in Education*, 44(3), 369–384. https://doi.org/10.1080/19415257.2017.1316299

Praetorius, A. K., Klieme, E., Herbert, B., & Pinger, P. (2018). Generic dimensions of teaching quality: The German framework of three basic dimensions. *ZDM Mathematics Education*, 50(3), 407–426. https:// doi.org/https://doi.org/10.1007/s11858-018-0918-4

Reuker, S. (2017). The noticing of physical education teachers: A comparison of groups with different expertise. *Physical Education and Sport Pedagogy*, 22(2), 150–170. https://doi.org/10.1080/17408989.2016.1157574

Rigdon, E. E. (1996). CFI versus RMSEA: A comparison of two fit indexes for structural equation modeling. *Structural Equation Modelling: A Multidisciplinary Journal*, 3(4), 369–379. https://doi.org/10.1080/10705519609540052

Sabers, D. S., Cushing, K. S., & Berliner, D. C. (1991). Differences among teachers in a task characterized by simultaneity, multidimensional, and immediacy. *American Educational Research Journal*, 28(1), 63–88. https://doi.org/10.3102/00028312028001063.(28(1))

Santagata, R., König, J., Scheiner, T., Nguyen, H., Adleff, A.-K., Yang, X., & Kaiser, G. (2021). Mathematics teacher learning to notice: A systematic review of studies of video-supported teacher education. *ZDM Mathematics Education*, 53(1), 119–134.

Schack, E. O., Fisher, M. H., & Wilhelm, J. (2017). *Research in mathematics education. teacher noticing: Bridging and broadening perspectives, contexts, and frameworks*. Springer.

Scheiner, T. (2021). Towards a more comprehensive model of teacher noticing. *ZDM Mathematics Education*, 53(1), 85–94. https://doi.org/10.1007/s11858-020-01202-5

Schlesinger, L., Jentsch, A., Kaiser, G., König, J., & Blömeke, S. (2018). Subject-specific characteristics of instructional quality in mathematics education. *ZDM Mathematics Education*, 50(3), 475–490.

Seidel, T., & Stürmer, K. (2014). Modeling and measuring the structure of professional vision in preservice teachers. *American Educational Research Journal*, 51(4), 739–771. https://doi.org/10.3102/0002831214531321

Sherin, M. G., Jacobs, V. R., & Philipp, R. A. (2011a). *Studies in mathematical thinking and learning: Mathematics teacher noticing*. Routledge.

Sherin, M. G., Jacobs, V. R., & Philipp, R. A. (2011b). Situating the study of teacher noticing. In M. G. Sherin, V. R. Jacobs, & R. A. Philipp (Eds.), *Studies in mathematical thinking and learning: Mathematics teacher noticing* (pp. 3–13). Routledge.

Sherin, M. G., & van Es, E. A. (2009). Effects of video club participation on teachers' professional vision. *Journal of Teacher Education*, 60, 20–37.

Shulman, L. (1986). Those who understand: knowledge growth in teaching. *Educational Researcher*, 15(2), 4–14.

Stahnke, R., Schueler, S., & Roesken-Winter, B. (2016). Teachers' perception, interpretation, and decision-making: A systematic review of empirical mathematics education research. *ZDM Mathematics Education*, 48(1–2), 1–27. https://doi.org/10.1007/s11858-016-0775-y

Steffensky, M., Gold, B., Holdynski, M., & Möller, K. (2015). Professional vision of classroom management and learning support in science classrooms—Does professional vision differ across general and content-specific classroom interactions? *International Journal of Science and Mathematics Education*, 13(2), 351–368. https://doi.org/10.1007/s10763-014-9607-0

Stigler, J. W., & Miller, K. F. (2018). Expertise and expert performance in teaching. In K. A. Ericsson, R.

R. Hoffman, A. Kozbelt, & M. A. Williams (Eds.), *The Cambridge handbook of expertise and expert performance* (2nd ed.pp. 431–452). Cambridge University Press. https://doi.org/10.1017/9781316480748.024

Stockero, S. L. (2021). Transferability of teacher noticing. *ZDM Mathematics Education*, 53(1), 73–84. https://doi.org/10.1007/s11858-020-01198-y

Stockero, S. L., & Van Zoest, L. R. (2013). Characterizing pivotal teaching moments in beginning mathematics teachers' practice. *Journal of Mathematics Teacher Education*, 16, 125147.

Tatto, M.T., Schwille, J., Senk, S., Ingvarson, L., Rowley, G., Peck, R., Bankov, K., Rodriguez, M. & Reckase, M. (2012). *Policy, practice, and readiness to teach primary and secondary mathematics in 17 countries. Findings from the IEA Teacher Education and Development Study in Mathematics (TEDS-M).* IEA.

van Es, E. A. (2011). A framework for learning to notice student thinking. In M. G. Sherin, V. R. Jacobs, & R. A. Philipp (Hrsg.), *Studies in mathematical thinking and learning. Mathematics teacher noticing* (pp. 134–151). Routledge.

van Es, E. A., & Sherin, M. G. (2002). Learning to notice: Scaffolding new teachers' interpretations of classroom interactions. *Journal of Technology and Teacher Education*, 10, 571–596.

van Es, E. A., & Sherin, M. G. (2021). Expanding on prior conceptualizations of teacher noticing. *ZDM Mathematics Education*, 53(1), 17–27. https://doi.org/10.1007/s11858-020-01211-4

Wu, M. L., Adams, R. J., & Wilson, M. R. (1997). *ConQuest: Multi-aspect test software* (Version 4.5.2) [computer program]. Australian Council for Educational Research.

Yang, X., Kaiser, G., König, J. & Blömeke, S. (2021a). Relationship between Chinese mathematics teachers' knowledge and their professional noticing. *International Journal of Science and Mathematics Education*, 19, 815–837. https://doi.org/10.1007/s10763-020-10089-3

Yang, X., König, J.,& Kaiser, G. (2021b). Growth of professional noticing of mathematics teachers: A comparative study of Chinese teachers noticing with different teaching experience. *ZDM Mathematics Education*, 53(1), 29–42. https://doi.org/10.1007/s11858-020-01217-y

Publication II

5

In 2023, the second publication on which this thesis is based, "The link between expertise, the cognitive demands of teacher noticing and, experience in teaching mathematics in secondary schools," was published in the International Journal of Science and Mathematics Education. Following on from the first publication and the evidence presented there for a stagnation, and perhaps even a decline, in teacher noticing skills from early career teachers to experienced teachers, this publication examined the demand-related differences in teacher noticing between the three groups that were the focus of Publication I. It investigated the differences in teacher noticing between groups with different lengths of teaching experience and suggested possible reasons for these differences. Since the research presented in this publication also examined cognitive demands, which were defined as competence components that refer to the knowledge domains needed to apply teacher noticing in the classroom, the publication is relevant not only to teaching experience and situation-specific skills but also to knowledge and cognitive processes in the expanded impact model (see Fig. 5.1).

In the paper, we investigated what differences exist between the noticing skills of three groups of teachers with different lengths of teaching experience in dealing with the cognitive demands of applying teacher noticing in certain teaching situations, as measured by a teacher noticing instrument related to those situations. For example, one of the demands investigated was the handling of situations that required knowledge of recent (mathematics) pedagogical topics. With its research interest, the study was directly related to the first publication (see Chap. 4) and the differences in the teacher noticing facets presented there. By introducing the concept of cognitive demands and a newly developed category system for analyzing the items of teacher noticing measures, one aim was to

A. Bastian, *Teacher Noticing of Pre-service and In-service Secondary Mathematics Teachers*, Perspektiven der Mathematikdidaktik,
https://doi.org/10.1007/978-3-658-45619-1_5

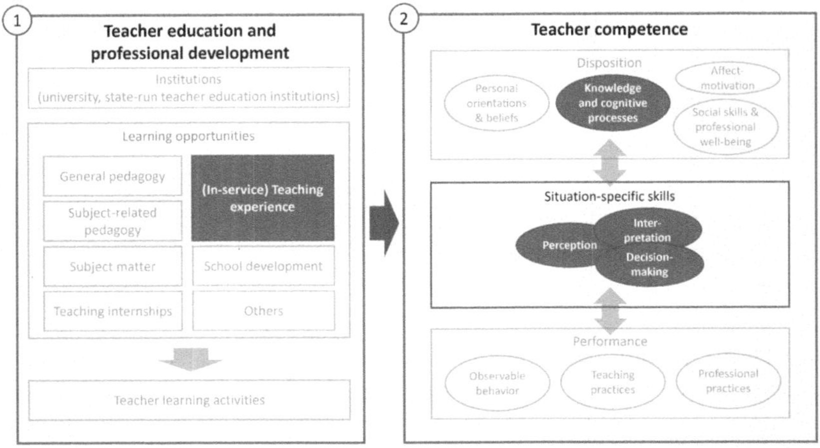

Note. **Areas examined by the research presented in Publication II are marked in blue.**

Fig. 5.1 Excerpt of the Expanded Impact Model to Illustrate the Contribution of Publication II

differentiate the differences between the investigated groups and to relate them to certain parts of professional knowledge. To address this research aim and question, the sample of 457 pre-service and in-service teachers analyzed in the first publication was re-examined with a new research lens. In this paper, we developed a new category system to match the items of the TEDS teacher noticing instrument to the cognitive demands that they place on teachers. We then formed scales of these categories using Rasch models and conducted multivariate analyses of variance and covariance to examine group influences on the handling of the cognitive demand categories.

The results suggested a significant impact of group membership on the management of cognitive demands with a medium effect size. Group comparisons between master's students and early career teachers revealed significant differences with varying effect sizes for all cognitive demands, particularly for the mathematics pedagogical perspective. Moreover, we contrasted early career teachers and experienced teachers to investigate the results that suggested advantages for the former group in Publication I. In the sample, early career teachers outperformed experienced teachers in three areas: decision-making skills, mathematics pedagogical knowledge, and cognitive demands related to recent mathematics pedagogical and general pedagogical topics. The latter findings in particular emphasized the need for effective professional development programs for experienced teachers in these areas.

5.1 Individual Contribution Statement

Gabriele Kaiser, Johannes König, Dennis Sommer, and I collaborated on the research project reported in Publication II. I proposed the research interest to follow up on the results of Publication I and conducted the item analyses to create a category system describing the cognitive demands of the items of our test instrument and the corresponding teaching situations. Following this, I conducted all statistical analyses on the sample, including Rasch scaling of the newly formed item categories and analyses of variance and covariance, and wrote a first draft of the entire paper. Subsequently, I revised the paper based on feedback from all co-authors. Gabriele Kaiser submitted the manuscript to the journal. All revisions based on the two rounds of reviews were made by me in close collaboration with Gabriele Kaiser and Johannes König. In addition, I presented the research in different stages at the summer schools for young researchers of the European Society for Research in Mathematics Education in 2021 and 2022.

5.2 Publication Imprint

The following manuscript was published originally by Springer Nature 2024 in the International Journal of Science and Mathematics Education in 2024[1]. It is reprinted here under the terms of the Creative Commons CC BY license.

The Link Between Expertise, the Cognitive Demands of Teacher Noticing and, Experience in Teaching Mathematics in Secondary Schools

Anton Bastian[1] (ORCID: 0000-0002-1177-6336), Gabriele Kaiser[2,*] (ORCID: 0000-0002-6239-0169)[1,4*], Dennis Meyer[3] (ORDCID: 0000-0003-1608-124X), Johannes König[4] (ORCID: 0000-0003-3374-9408)

[1]University of Hamburg, Von-Melle-Park 8, 20146 Hamburg, Germany

---

[1] Bastian, A., Kaiser, G., Meyer, D., & König, J. (2024). The link between expertise, the cognitive demands of teacher noticing and, experience in teaching mathematics in secondary schools. International Journal of Science and Mathematics Education, 22, 257-282. https://doi.org/10.1007/s10763-023-10374-x

[2]University of Hamburg / Nord University, Von-Melle-Park 8, 20146 Hamburg, Germany
[3]University of Hamburg (ProfaLe), Bogenallee 11, 20144 Hamburg, Germany
[4]University of Cologne, Gronewaldstraße 2a, 50931 Köln, Germany
*Corresponding author: Gabriele Kaiser, gabriele.kaiser@uni-hamburg.de

Abstract

Teacher noticing, a situation-specific part of teachers' professional competence, has gained much importance in the past two decades. The construct is believed to play a crucial role in the transfer of knowledge and skills into professional teaching performance. However, there is little empirical evidence on how the length of teachers' teaching experience affects the development of expertise in teacher noticing. In particular, little research has been conducted on the development of facets of teacher noticing—namely perception, interpretation, and decision-making—as well as the handling of cognitive demands while applying teacher noticing. To fill this research gap, the present study investigates the teacher noticing of master's students, early-career teachers, and experienced teachers in the context of secondary mathematics via cross-sectional comparison. Teacher noticing skills and differences between the three groups are analyzed based on participants' ability to deal with the cognitive demands of an established video-based teacher noticing instrument. The results suggest significant increases in teacher noticing from students to in-service teachers, especially in mathematic-specific areas. Hardly any significant differences were found between early-career and experienced teachers, which suggests a stagnation in teacher noticing skills. However, the early-career teachers showed better performance in decision-making and mathematics-related items as well as knowledge-based items related to recent topics from mathematics pedagogy and general pedagogy. This result emphasizes the importance of knowledge prerequisites for the development of expertise in teacher noticing. It also highlights that expertise development is not directly connected to the length of teaching experience.

Keywords: Teacher noticing · Teaching experience · Mathematics secondary pre-service and in-service teachers · Cognitive demands · Expert–novice comparisons · Item analysis

5.2.1 Introduction

Due to teachers' significant impact on students' learning, teacher education represents a main focus of education policy and society. Teachers' professional competencies and their development have been primary concerns in educational—particularly mathematics pedagogical—research in many parts of the world for over two decades, as demonstrated by seminal large-scale studies in this field (Blömeke et al., 2010; Kunter et al., 2011). In recent years, situation-specific competence facets, especially teacher noticing, which is a construct mediating between dispositions and teaching performance, have increasingly come into focus (Blömeke et al., 2015). The construct of teacher noticing describes teachers' ability to selectively perceive relevant information in complex teaching situations, to analyze competently, and thus to act appropriately (Jacobs et al., 2010; Kaiser et al., 2017). Therefore, teacher noticing is a central component of teachers' cognition, an important characteristic of teaching expertise, and a crucial factor affecting students' learning success (Blömeke et al., 2022; Lachner et al., 2016; Yang et al., 2021).

As a consequence, understanding the construct and developing pre-service and in-service teacher's noticing skills are central goals of mathematics pedagogical research (Amador et al., 2021; Dindyal et al., 2021). However, to date only few studies have investigated the development of expertise in teacher noticing and have considered the influences of teaching experience on this development (König et al., 2022). In particular, there are no empirical studies addressing how the cognitive demands of teacher noticing depend on teaching experience (Bastian et al., 2022; Yang et al., 2021). Classroom situations, tasks, and problems, as well as items of the teacher noticing instrument that refer to them, impose cognitive demands on teachers (for a discussion on cognitive demands, see Turner et al., 2015).

Due to the limited number of studies examining the cognitive demands of teacher noticing—imposed by the noticed event, related situation, or accompanying items in the evaluation—this study focuses on demand-related differences in teacher noticing between three groups with different lengths of teaching experience. A detailed category system to describe the cognitive demands that mathematics teachers face during noticing was developed, and a comparison of the experience groups' handling of different cognitive demands was conducted.

5.2.2 Theoretical Framework and State of the Art

Professional noticing of classroom events as a core activity of teachers entails perceiving (instructionally) relevant events in demanding situations and then processing this information (Sherin et al., 2011; van Es & Sherin, 2002). The construct of teacher noticing can be generally understood as a set of skills that enables teachers to deal with the complexity of teaching situations and to act professionally within them (Dindyal et al., 2021; Sherin et al., 2011). Since the early 2000s, exploration of this construct has become increasingly important in the educational sciences, especially in mathematics pedagogical research (see overviews by Amador et al., 2021; Dindyal et al., 2021; König et al., 2022; Schack et al., 2017). Moreover, teacher noticing has been shown to be an essential factor for teachers' professional competence, teachers' performant behavior in the classroom, and students' learning success (Blömeke et al., 2022; Kaiser et al., 2017; Santagata & Yeh, 2016).

Currently, there is no consensus on the structure and characteristics of teacher noticing (Dindyal et al., 2021; Yang et al., 2021), as different theoretical perspectives have influenced the academic discourse and conceptualization of the construct. For instance, in a recent systematic literature review, König et al. (2022) identified four influential perspectives and research traditions in the current research. Within the cognitive–psychological perspective, teacher noticing is understood as cognitive processes that take place mentally in the individual teacher. Protagonists include, among others, Jacobs et al. (2010), Seidel and Stürmer (2014), and van Es and Sherin (2002, 2021). This perspective can be considered the most influential one at present (König et al., 2022), and it forms the theoretical basis of this study. In their survey of theoretical frameworks for noticing, Amador et al. (2021) described two cognitive–psychological approaches, the learning to notice framework proposed by van Es (2011) and the professional noticing of children's mathematical thinking framework proposed by Jacobs et al. (2010), as predominant in the discussion and understanding of teacher noticing. Other perspectives that influence the discourse include the socio-cultural approach, which focuses on the social construction of professional vision[2] (Goodwin, 1994); the discipline-specific approach introduced by Mason (2002); and the expertise paradigm (e.g., Berliner, 1994).

Within the cognitive–psychological perspective, different facets of teacher noticing have been analytically distinguished: perceptual, interpretive, and, if applicable, decision-related skills (König et al., 2022). For example, Sherin et al. (2011, p. 5) described the facets "attending to particular events in an instructional setting" and "making sense of events in an instructional setting" as crucial for teacher noticing. In addition, a considerable number of recent studies included a decision-related facet in teacher noticing conceptualizations (Amador et al., 2021; König et al., 2022). For example, Jacobs et al. (2010, p. 172) identified the facets attending to children's strategies, interpreting children's understandings, and deciding how to respond on the basis of children's understandings. Likewise, our own framework includes a decision-related facet, namely decision-making (Kaiser et al., 2015).

As pointed out in the literature review performed by König et al. (2022), students' mathematical thinking and understanding are central topics in the research on teacher noticing. Some studies have investigated mathematical or mathematics pedagogical topics, such as representations (Dreher & Kuntze, 2015) or algebra (Walkoe, 2015). In addition, studies have examined teacher noticing in relation to general pedagogical issues, such as classroom management, instructional materials, and curricula (Dietiker et al., 2018; Gold & Holodynski, 2017; Stahnke & Blömeke, 2021) as well as social participation and equity (Louie et al., 2021).

5.2.2.1 Teacher Noticing as a Competence Facet

To complement the conceptualization of teachers' professional competence with situation-specific skills, teacher noticing has been included as a central component in competence frameworks (Kaiser et al., 2017). In their competence as continuum model, Blömeke et al. (2015) conceptualized perceiving, interpreting, and decision-making as the mediating components between teachers' dispositions and classroom performance. Thus, professional competence was understood neither solely as disposition nor performance, but as a horizontal continuum from one

[2] In the current discussion, the notions of teacher noticing and professional vision are used to describe strongly overlapping situation-specific constructs that mediate between disposition and performance (Martin et al., 2022). Although professional vision originates from a social–cultural understanding, it is, like teacher noticing, often used in cognitive–psychological approaches (Gegenfurtner, 2020). In this paper, we use teacher noticing to describe our theoretical framework but include research concerning professional vision (see König et al. (2022) for an overview).

person's knowledge and affective dispositions through situation-specific skills, which embed dispositions in context, to teachers' actual performance in the classroom. As a consequence, teacher noticing was understood as strongly dependent on teachers' dispositions, particularly knowledge facets, which aligns with prior research on teacher noticing and recent empirical studies (Blömeke et al., 2022; van Es & Sherin, 2002, 2021).

The competence as continuum model was further developed in other conceptualizations. For example, Santagata and Yeh (2016) emphasized the bidirectional relationship between dispositions and situational skills, which contrasts the rather linear understanding of the competence as continuum model. In their recent work, Metsäpelto et al. (2021) developed this model into a multidimensional adapted process model of teaching, which combines teacher noticing, as a situation-specific skill, with teaching practices and professional practices under the term "teaching competences." In addition, this model enriches dispositions with cognitive and social skills and includes students' learning success by looking at teachers' effectiveness. Nevertheless, teacher noticing remains a central point in this conceptualization of competence (Metsäpelto et al., 2021).

Empirical studies have highlighted the relevance of situation-specific skills in the development of teachers' professional competence, and have supported corresponding models (Santagata & Yeh, 2016). Previous studies have found significant correlations between teacher noticing and knowledge (e.g., Blömeke et al., 2016; Dreher & Kuntze, 2015; Meschede et al., 2017), belief facets (e.g., Meschede et al., 2017), and instruction (e.g., Blömeke et al., 2022), as well as the construct's empirical separability from other competence facets (Bastian et al., 2022; Blömeke et al., 2016; Copur-Gencturk & Tolar, 2022), which justify its integration into these competence models.

5.2.2.2 Expertise Development in Teacher Noticing and the Impact of Teaching Experience

Thus far, only a few empirical studies have investigated teacher noticing in the context of expertise and the impact of teaching experience (König et al., 2022; Yang et al., 2021). A seminal exception is the American study by Jacobs et al. (2010), in which four different groups of primary teachers with varying levels of expertise were cross-sectionally compared. The results of the study demonstrated that skills in teacher noticing significantly increased across the four expertise groups. The evidence suggests that teaching experience as well as professional development, another expertise characteristic used in this study, influenced teacher noticing. There was a strong increase in noticing skills from the group of pre-service teachers to the group of in-service teachers, especially in the noticing facets of perception and interpretation; apparently, these facets increase with teaching experience. In contrast, expertise in the decision-related facet seemed to develop mainly in combination with professional development (Jacobs et al., 2010). The significant increase in expertise in teacher noticing between pre-service and in-service teachers, and thus the substantial influence of teaching experience on teacher noticing, was confirmed in further quantitative studies (Gold & Holodynski, 2017; König & Kramer, 2015; Meschede et al., 2017; Yang et al., 2021). However, few studies have investigated the reasons for these findings or the areas of development in detail.

Extensive research from the German context have demonstrated that master's students (MS) have significantly lower skills in all three facets of teacher noticing compared to early-career teachers (ECT) and experienced teachers (ET) (Bastian et al., 2022), but no significant differences were found between the ECT and ET, except for a slight advantage for the ECT in the decision-making facet. However, the ET scored nominally lower on average in all three noticing facets, providing evidence of a stagnation—or even decline—in teacher noticing expertise. A similar finding was demonstrated by Kleickmann et al. (2013) for teachers' professional knowledge. Additional regression analyses showed significant negative slopes in teacher noticing with increasing length of teaching experience (Bastian et al., 2022). These results contrast the analyses performed by Jacobs et al. (2010), which are described above, and East Asian studies, in which the development of mathematics teachers' noticing skills progressed nearly linearly over the groups of MS, ECT, and ET (Yang et al., 2021). These differences indicate that further research is needed to explore differences in noticing skills between these groups of teachers and examine the stagnation or slight decline of noticing skills among ETs in the Western context. Both topics will be addressed in this study.

Recently, innovative techniques, such as eye-tracking procedures, have been used to investigate noticing. Studies using these techniques have reported that expert teachers focus on areas of teaching situations that contain information relevant to instruction, whereas novices let their gaze wander more frequently (Wolff et al., 2016). Moreover, the perception of experts seemed to be more knowledge-based than that of novices (Gegenfurtner et al., 2020; Seidel et al., 2021; Wolff et al., 2016). Novices observing classroom interactions were often solely focused on teachers, while experts concentrated on students and observed events related to instruction units, curricula, or school culture (Gegenfurtner et al., 2020; McDonald, 2016). Further, experts diagnosed more accurately, interpreted more often, detailed what they saw, and offered more alternative actions for observed lessons (Seidel et al., 2021; Stahnke & Blömeke, 2021).

Expertise research has pointed out that expertise must be conceptualized as domain-specific and confined to particular contexts (Berliner, 2001; Boshuizen et al., 2020), which has been observed for expertise in teacher noticing. For example, studies by Yang et al. (2021) in an East-Asian context suggested that ET demonstrate expertise in dealing with teaching-related reflections, student thinking, and assessment and diagnosis of mathematical content. In contrast, ECT excel in topics related to student participation in the classroom, cooperative teaching methods, and more recent mathematics pedagogical topics, such as the teaching of mathematical modelling.

5.2.2.3 Conceptualization and Theoretical Framework in the TEDS-M Research Program

Situation-specific components have been incorporated into the measurement of teachers' competence. In the Teacher Education and Development Study in Mathematics (TEDS-M) research program, teacher noticing was conceptualized as a part of the competence framework and has been included in competence measurement since the TEDS-Follow-Up (TEDS-FU) study (Blömeke et al., 2014; Kaiser et al., 2015). Referring to the competence as continuum model proposed by Blömeke et al. (2015), teacher noticing is understood here as situation-specific skills and is conceptualized as consisting of three noticing facets:

(a) perceiving particular events in an instructional setting; (b) interpreting the perceived activities in the instructional setting; (c) decision-making, either as anticipating responses to students' activities or as proposing alternative instructional strategies. (Kaiser et al., 2015).

This analytical, cognitive–psychological definition has perceptual, interpretive, and decision-related facets. With reference to the expertise paradigm (Berliner, 1994; Carter et al., 1988), the first facet describes a perceptual process in which one observes something without or with only minimal interpretation (Bastian et al., 2022), i.e., identifying discernable events. Interpretation of these incidents is understood as analysis based on the person's knowledge and beliefs and connection of new knowledge chunks with existing knowledge or each other. The third facet of the conceptualization explicitly includes decision-making as a significant component of teacher noticing. In line with the work of Jacobs et al. (2010), decision-making is defined as the meaningful construction of possible continuations of the lesson or instructional incident, the formulation of reactions to student behavior, or the offering of well-founded alternatives for observed teacher acts (Yang et al., 2021).

Thematically, this understanding of teacher noticing does not focus on single mathematical topics. Instead, it includes a broad field of teaching events and conditions that are relevant for high-quality mathematics education. These include, among others, the cognitive activation of students, the design of teaching–learning processes, and effective classroom management. In comprehensive descriptions of the construct, each facet is understood from both a general pedagogical and mathematics pedagogical perspective (Kaiser et al., 2015; Yang et al., 2021).

5.2.2.4 Research Question

Based on the current state of the art presented above and identified research gaps— particularly the contradicting results concerning teachers' development of expertise in teacher noticing with increasing length of teaching experience— this study aims to investigate the teacher noticing skills of different teaching experience groups. It goes beyond statistical comparisons of average scores in teacher noticing facets and takes a closer look at the in-depth differences among those experience groups in their handling of different cognitive demands. The following research question is examined:

> What differences exist between the teacher noticing of groups with different levels of teaching experience in managing the cognitive demands of teacher noticing measures and related teaching situations?[3]

We followed the generally agreed assumption that teaching experience facilitates higher expertise levels in teacher noticing, giving advantages to in-service against pre-service teachers. These advantages of length of teaching experience may vary in effect size for differing cognitive demands and different experience groups. We therefore assume that ECT might be more competent in dealing with recent mathematics pedagogical topics (Yang et al., 2021).

5.2.3 Methodological Approach

The sample, test instrument, and item and data analysis are described below. Supplementary details are provided in the electronic supplemental materials (ESM).

[3] The cognitive demands mentioned arise from the items which are included in the teacher noticing measure and refer to certain teaching situations (Weyers et al., 2023).

5.2.3.1 Study Context and Sample

The sample consists of N = 457 secondary mathematics pre-service and in-service teachers, who participated in one of four studies from the TEDS-M research program (see Table 5.1).

Table 5.1 TEDS-M studies included in the sample

Study	Survey period	Participants
TEDS-Follow Up	2011–2013	Early-career teachers from Germany who were former participants of the TEDS-M study
TEDS-Instruct	2014–2016	In-service teachers from the federal state of Hamburg, Germany
TEDS-Validate	2016–2019	In-service teachers from the federal states of Thuringia, Saxony, and Hesse, Germany
TEDS-Validate-Transfer	2020–2021	Master's students in mathematics education from six German universities

The participants were divided according to their length of teaching experience: MS (n = 110) without extensive and systematic teaching experience, ECT (n = 193) with an average of 4.6 years of in-service teaching experience (SD = 0.5, Min = 3.5, Max = 6.0), and ET (n = 154) with an average of 19.6 years of in-service teaching experience (SD = 10.4, Min = 6.5, Max = 41.5). As teaching experience is a necessary—although not solely sufficient—prerequisite for expertise in the teaching profession (Caspari-Sadeghi & König, 2018; Palmer et al., 2005), we intend to explore the influence of teaching experience on the development of expertise in noticing. The three groups represent three different stages of teaching experience; thus, we hypothesize that they are sufficiently different in light of the expertise paradigm (Berliner, 1994; Gruber et al., 2019; Stigler & Miller, 2018). As the data collection took place at different times for the three expertise groups—first the ECT were tested, then ET, then MS, differences in noticing may be caused by differences in the educational background of the three groups. We will come back to this aspect in the discussion and interpretation of the results.

Further characterizations of the three experience groups can be found in Table 5.2. The groups show no significant differences in terms of gender, diploma from German secondary school qualifying for university admission or matriculation (so-called Abitur) and relevant teacher education grades. However, there are significant differences between the groups with regard to school type, which different selectivity—meaning the different socio-economic background of the students and their orientation in academic track and non-academic track schools—is of high importance in the German school system. This will be considered in the results section.

Table 5.2 Descriptive characterization of the total sample and the experience groups

Variable					Experience groups				Test for Group differences	
	Total		MS		ECT		ET		χ2 (df)/ odds ratio/t	*p*
Sample size	457		110		193		154		---	---
Gender (% female)	59.0		66.4		58.3		54.5		3.76 (2)	.152
School type (% academic track)	48.0		40.0		52.3		45.5		6.72 (2)	.035
	M	*SD*	*M*	*SD*	*M*	*SD*	*M*	*SD*		
Age (in years)	34.8	10.9	24.6	4.08	31.7	5.5	46.0	9.5	488.36	< .001
Teaching experience (in years)	9.8	10.2	0.0	.00	4.6	.50	19.6	10.4	397.45	< .001
Grade[a] (Abitur[b])	2.1	.62	2.1	.63	2.1	.61	---	---	.24 (222)	.813
Grade[a] (University)	1.9	.57	---	---	1.9	.54	2.0	.61	1.18 (304)	.239
Grade[a] (induction phase[c])	2.0	.65	---	---	2.0	.61	2.0	.67	.11 (158)	.913

Note. MS ≙ master's students, ECT ≙ early-career teachers, ET ≙ experienced teachers. To test for group differences, chi-square tests or Fisher's exact tests were used when comparing three groups, and two-tailed t-tests were used when comparing two groups.

[a] German grades range from 1 (best) to 4 (passed).

[b] German school leaving diploma, that qualifies for university attendance.

[c] Second phase of formal teacher education, which focuses on practical teacher education accompanying teaching at school.

5.2.3.2 TEDS-M Instrument for Measuring Teacher Noticing

The TEDS-M instrument for measuring teacher noticing was developed in the TEDS-FU study. It is an established instrument for measuring teacher noticing as a contributor to teacher expertise. We use the achievements in this test as a norm- or task-oriented measure of expertise to identify expert teachers (Krauss, 2020; Stigler & Miller, 2018). The evaluation of noticing as a facet of teachers' professional competence is based on three scripted (i.e., staged) video-vignettes with lengths of 2.25 to 3.5 min. The entire test instrument consists of three randomly arranged test units and has a duration of approximately 90 min. In each test unit, the participants first received background information regarding the class shown in the video-vignette, the general teaching conditions, and the mathematical content covered. Afterward, they received one-time access to the corresponding video-vignette. This allowed the teaching situation to be observed as realistically as possible in the test environment. Overall, the use of scripted video-vignettes instead of filmed natural instruction or text-vignettes enabled a manageable, cognition-activating, and strongly classroom-related measurement (Kramer et al., 2020; Piwowar et al., 2018; Santagata et al., 2021). The three video-vignettes contained a range of different mathematical topics (e.g., surface and volume calculation, functions, modeling) and teaching phases like the introduction of a mathematical task and work on it or a plenary with discussion of results (for further description, see Kaiser et al., 2015). They represented lessons in the 9th to 10th grades at different school types (Kaiser et al., 2015). After watching the video-vignette, the participants answered open-response and Likert-type rating scale items that addressed their abilities in one of the three teacher noticing facets with a mathematics pedagogical or general pedagogical focus (Blömeke et al., 2014, see Table 5.3 for the respective number of items). Across all three video-vignettes, the measurement instrument included a total of 77 items.

As displayed in Table 5.3, the pedagogical perspective on teacher noticing (P_ PID) is well represented in the item set measuring perception, while decision-making items particularly comprise items concerning the mathematics pedagogical perspective (M_PID), which was a result of scaling and test construction. For the interpretation facet, items are more or less equally distributed in terms of subject-related perspectives.

Concerning the item formats, a rating scale item consists of a statement and a four-point Likert scale, with which the statement is to be assessed. The example item in Fig. 5.2 requires perception of a specific classroom event.

Example items for the open-response format and the interpretation and decision-making facets are displayed in Figs. 5.3 and 5.4. The interpretation item

Table 5.3 Number of items in the noticing measure by noticing facet and subject-related perspective

Noticing Facet	Subject-related perspective		Total
	M_PID items	P_PID items	
Perception	3	21	24
Interpretation	25	17	42
Decision-making	8	3	11
Total	36	41	77

Note. M_PID ≙ Mathematics Instruction: Perception, Interpretation, Decision, P_PID ≙ Pedagogy: Perception, Interpretation, Decision.

The guiding question of the lesson is presented in written form.	fully correct	partially correct	partially incorrect	not correct at al
	o	o	o	o

Fig. 5.2 Rating scale item for the perception facet (P_PID)

asks participants to analyze a student's way of solving a volume and surface calculation and identify indicators of a formal mathematical approach. Consequently, this item is related to M_PID. The decision-making item concerning P_PID asks for possible changes to the course of instruction to better deal with the class's heterogeneity.

What indicators in the video support the assumption that Kim prefers a formal mathematical approach rather than a descriptive one?

Name three.

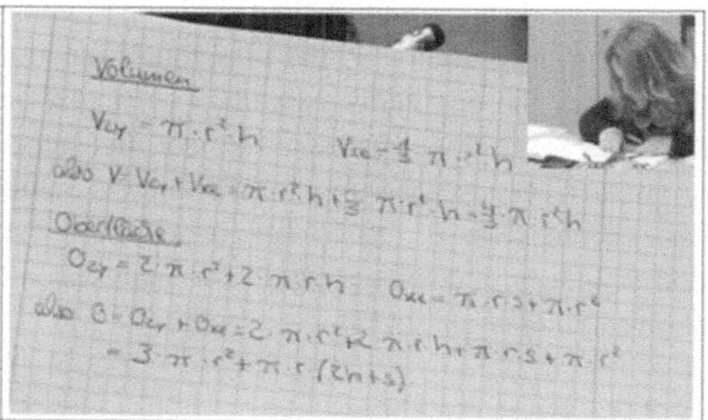

Fig. 5.3 Open-response item concerning the interpretation facet (M_PID)

Assume that the teacher feels that he is not addressing the ability heterogeneity of his class shown in the video with his instruction adequately.

Describe briefly,
a) at which point in the course of instruction
b) and in which way
you would modify the lesson shown - while retaining as many of its elements as possible - in order to better accommodate the different learning conditions within the class.

Fig. 5.4 Open-response item concerning the decision-making facet (P_PID)

Responses to the rating scale items were dichotomously coded as incorrect or correct based on intensive expert interviews (Hoth et al., 2016). In these interviews, experts answered the items and had the opportunity to evaluate and comment them. Items with at least 60% of agreement on one of the four possible answers were included in the test with the agreed answer as the correct choice. Items with a minimum of 80% of agreement for one tendency (approval or disapproval for the given statement) were revised considering the experts' comments and then again reviewed. To evaluate answers to open-response items, a coding manual was developed based on theoretical considerations and discussions with experts. The manual provided detailed descriptions and numerous anchor examples to illustrate which answers are to be considered correct. To test the reliability of this procedure, 20% of the answers to each item were double-coded, and the intercoder reliability was examined based on Cohen's kappa (Cohen, 1960). Good overall agreement ($\kappa_{mean} = 0.85$) was achieved (see ESM for details). The validity of the measurement instrument was ensured through various procedures, e.g., extensive workshops with mathematics pedagogy and general pedagogy experts concerning authenticity of the classroom situations in the video-vignettes and adequacy of the test items as well as curricular analyses of the covered content (Kaiser et al., 2015). The measurement of teacher noticing was independent of the video-vignettes; the items for the three vignettes measure the underlying construct (Blömeke et al., 2015). An adaptation of the instrument in China achieved good fit values in confirmatory factor analyses, which confirms its cross-cultural validity (Yang et al., 2018).

5.2.3.3 Item Analyses

We developed a category system to perform a detailed comparison of the abilities of teacher groups with different lengths of teaching experience in managing different cognitive demands related to teacher noticing and its facets. Cognitive demands were defined as competence components relating to knowledge areas or abilities that are needed to apply teacher noticing and its facets in certain classroom situations as well as to answer items on the noticing measures referring to these situations (Turner et al., 2015). To develop the categories, qualitative text analysis and rational task analysis were carried out on all 77 items (Kuckartz, 2014; Resnick, 1975), which resulted in three different perspectives on cognitive demands: knowledge (5 categories), practice-oriented competence facets (5 categories), and additional characteristics (2 categories).

Relevant literature on the topics of teachers' competence domains, cognitive processes, and particularly knowledge served as the basis for the deductive category development. The conceptualization of competence, which we generally understood as the available or learnable set of cognitive skills and abilities to successfully handle a complex demand as well as the motivational, volitional and social willingness to do this (Weinert, 2001), has been regarded as a particularly relevant component of item analyses (Turner et al., 2015). To ensure theoretical as well as practical perspectives are considered, a combination of research-oriented and practice-oriented frameworks of teacher competence was applied. To include multiple aspects of the research-oriented approach of competence as continuum model (Blömeke et al., 2015), categories specifying knowledge-based aspects of noticing were developed in addition to the existing teacher noticing facet categories and subject-related perspectives (Blömeke et al., 2010). Since the situation-specific perspectives P_PID and M_PID are strongly connected to general pedagogical knowledge and mathematics pedagogical content knowledge, categories assigning the corresponding items to specific knowledge areas were developed (e.g., mathematics pedagogical curricula and planning or students' learning). Curricular analyses of competence frameworks for various parts of teacher education were included as practice-oriented competence frameworks, which led to categories such as dealing with heterogeneity.

Inductive categorization with an experienced coder led to additional categories concerning general pedagogical and mathematics pedagogical topics, which can be described as new in the current academic discourse. An excerpt of the category system with the number of items which were allocated to this category is shown in Table 5.4. Due to space restrictions, the complete category system can only be found in the ESM.

The strictly rule-guided procedure and creation of a detailed coding manual ensured internal study quality and internal validity (Kuckartz, 2014). With the help of two experienced coders, the coding manual was revised and inductively sharpened using the collected data. After double-coding 40% of all items, substantially significant intercoder reliability could be achieved for all items ($\kappa_{\text{Mean}} = .833$, $\kappa_{\text{Min}} = .684$, $\kappa_{\text{Max}} = 1.000$; Landis & Koch, 1977). Content validity

Table 5.4 Excerpt from the category system and number of occurrences

Categories	n
Teacher noticing facet	
Perception	24
Interpretation	42
Decision-making	11
Subject-related perspective	
P_PID	41
M_PID	36
Knowledge of mathematics pedagogy	
Not applicable	41
Curriculum & planning	16
Interaction	20
Knowledge of general pedagogy	
Not applicable	36
Students' learning	16
Teaching–learning processes	21
Students' assessment	4
Dealing with heterogeneity	
Not necessary	42
Necessary	24
Particularly necessary	11
Particular relevance of recent (mathematics) pedagogical topics	
Nonexistent	58
Existent	19

Note. For each category, each item was allocated to one of the subcategories. *n* represents the number of items in this subcategory in the TEDS-FU instrument for teacher noticing. Note that the number of occurrences for all subcategories of one category must add to the total number of items in the instrument. In total, the instrument comprised 77 items.

was established through the integration of existing theories, e.g., several existing competence frameworks, and methods, e.g., rational task analysis, related to item demand analysis. External quality criteria regarding transferability were met by involving external experts and comparing the results with existing measurement instruments.

5.2.3.4 Category System and Occurrence

Given the comprehensiveness of the category system, the presented analysis is limited to a few particularly interesting and relevant categories (see the ESM for the entire category system). A short description of these categories is given in Table 5.5.

The absolute frequencies of the occurrence of cognitive demand categories are displayed in Tables 5.3 and 5.4. Knowledge of mathematics pedagogy is divided more or less equally between curricula and planning (20.8%) and interaction (26.8%). Similar results are found for students' learning (20.8%) and teaching–learning processes (27.3%) for knowledge of pedagogy. Dealing with heterogeneity, which is an important topic on the global pedagogical agenda (e.g., König et al., 2017), is necessary in 31.2% of the items and particularly necessary in 14.3%. Consequently, this essential teacher skill is represented by the instrument. Knowledge of the particular relevance of recent (mathematics) pedagogical topics is needed in 37.7% of items. Standard correlation measures—Cramer's V and Spearman's rank correlation—were used to examine the correlations between categories. Mostly moderate correlations were observed, suggesting that distinct but correlated categories describe different aspects of cognitive demands (see ESM Table 2).

5.2.3.5 Scaling and Quantitative Data Analysis

To examine the relationships between the handling of cognitive demands and experience group membership, scales for each subcategory were developed and validated using Rasch models. These models achieved good model fit (for one factor model, RMSEA = 0.027). As an indicator for measurement invariance, item estimates were calculated for each experience group and then correlated (cf. Bond et al., 2021; König et al., 2017). These analyses indicated not perfect but acceptable measurement invariance ($r_{\text{MS-ECT}} = 0.81$, $r_{\text{MS-ET}} = 0.77$, $r_{\text{ECT-ET}} = 0.92$) when analyzing group differences.

The extent to which experience group membership explains variance with respect to the cognitive demands of items was investigated by applying multivariate analyses of variance (MANOVAs) for each demand category and its

Table 5.5 Short description of categories and subcategories

Category label	Description of category and its subcategories
Teacher noticing facet	The teacher noticing facet or skill essential for item handling, divided into *perception*, *interpretation*, and *decision-making*.
Subject-related perspective	The most relevant aspect—and thus knowledge facet—for item handling considering the *mathematics pedagogical (M_PID)* and *general pedagogical perspective (P_PID)*.
Knowledge of mathematics pedagogy	The most relevant specific knowledge area in mathematics pedagogy for item handling. *Curricula & planning*: educational curriculum, especially competence-oriented teaching, and knowledge about the planning of mathematics instruction. *Interaction*: knowledge needed for interactions with students, like diagnostic knowledge and knowledge about students' errors. *Not applicable*: no mathematics pedagogical knowledge is needed. P_PID items are assigned to this category by default.
Knowledge of general pedagogy	The most relevant specific knowledge area in general pedagogy for item handling. *Students' learning*: knowledge about students and their individual learning processes, like cognitive dispositions and motivation. *Teaching-learning-processes*: teaching methods, instructional lesson structuring, classroom management. *Students' assessment*: performance assessment and evaluation. *Not applicable*: no general pedagogical knowledge is needed. M_PID items are assigned to this category by default.

(continued)

Table 5.5 (continued)

Category label	Description of category and its subcategories
Dealing with heterogeneity	The extent to which item handling requires one to deal with heterogeneity in classroom, like perceiving heterogeneity, supporting individual students, and differentiating tasks.
Particular relevant knowledge of recent (mathematics) pedagogical topic	Whether item handling requires knowledge about topics like competence-oriented teaching, modeling and real-world references, open learning, language acquisition, and heterogeneity.

Note. For each category, each item was allocated to one of the subcategories. For the perspective category and the two knowledge categories the most relevant aspect/knowledge area for the item handling was determined using item analysis taking the possible item solutions into account.

subcategories. Following the MANOVAs, post hoc tests were performed to identify significant differences between individual groups. In the case of variance homogeneity, the Scheffé post hoc test was used for post hoc analyses. The Scheffé test is a more conservative method compared to other post hoc tests, and it can be used in the case of unequal sizes of comparison groups. In the absence of homoscedasticity, the Games-Howell post hoc test was used.

5.2.4 Results

We first present the results concerning the overall effect of teaching experience group membership on the handling of cognitive demands. Then, we more closely compare and contrast consecutive groups.

5.2.4.1 Relationship Between Teaching Experience Group Membership and Cognitive Demands

Examples of mean scores for two cognitive demands by experience group are presented in Table 5.6. These scores can be considered exemplary for the other categories.

Overall, analysis of the mean values in individual demand categories replicates the pattern described by Bastian et al. (2022) for the overall construct of teacher noticing and its facets. In-service teachers perform significantly better than MS,

Table 5.6 Mean values according to experience group in example categories

Variable	Experience groups						Test of group differences
	MS ($n = 110$)		ECT $n = 193$)		ET ($n = 154$)		Post hoc test[a]
	M	*SD*	*M*	*SD*	*M*	*SD*	
Subject-related perspective							
M_PID	42.94	9.87	53.34	8.86	50.85	8.92	MS < ECT, MS < ET, ET < ECT
P_PID	45.14	11.56	52.21	8.88	50.70	8.95	MS < ECT, MS < ET
Knowledge of mathematics pedagogy							
Curriculum and planning	43.32	10.51	53.34	8.91	50.58	8.55	MS < ECT, MS < ET, ET < ECT
Interaction	44.23	9.32	52.59	9.43	50.88	9.55	MS < ECT, MS < ET

Note. Scores were standardized per category and transformed to $M = 50$ and $SD = 10$, MS ≙ master's students, ECT ≙ early-career teachers, ET ≙ experienced teachers, M_ PID ≙ Mathematics Instruction: Perception, Interpretation, Decision, P_PID ≙ Pedagogy: Perception, Interpretation, Decision.

[a] All given differences of the post hoc test are statistically significant ($p < .05$). See ESM for more statistical information.

while ECT and ET hardly differ from each other (see Table 5.6 and ESM). However, ECT are nominally better than ET, suggesting stagnation or even a slight decline in ability to deal with the respective cognitive demands. These differences between the two teacher groups with teaching experience were not significant in prior analyses except for the decision-making facet (Bastian et al., 2022). Thus, it is of great interest to investigate the extent to which significant differences exist when individual cognitive demands are considered. The newly developed category system facilitates these investigations.

MANOVAS were conducted to analyze whether experience group membership is a decisive factor affecting the handling of cognitive demands. As Table 5.7 illustrates, all MANOVAs, which are qualified by Wilks-λ, suggest that experience group membership is a significant factor affecting performance in dealing with every cognitive demand. Since the groups differed significantly in school type (see Table 5.2), this characteristic was included as a covariate in an additional analysis. However, the characteristic showed no significant influence and thus was not considered in further analyses. The significant variance explanation of group membership for individual demand categories (between 7.5% and

10.5%) can be classified as a medium effect (Cohen, 1988). This means that experience group membership can be attributed a medium practical significance for the handling of different cognitive demands. The largest effects are found for knowledge of pedagogy, with 10.5% variance explanation, followed by knowledge of mathematics pedagogy, with 9.4% variance explanation.

Deeper investigation of the variance explanations for individual subcategories reveals a more differentiated picture of small to large effects (see ESM). Interestingly, group membership has large effects for the demand categories of curriculum and planning in mathematics pedagogy ($\eta^2 = .156$), knowledge of pedagogy not applicable ($\eta^2 = .170$), the subject-related perspective M_PID ($\eta^2 = .170$), and the particular relevance of recent (mathematics) pedagogical topics existent ($\eta^2 = .149$). Thus, in combination of the large effects observed for items for which no knowledge of pedagogy is needed and items belonging to the M_PID subject-related perspective, a large effect of group membership can be found for mathematics-related items. This result explains the variance explanation of knowledge of pedagogy reported above, which is probably not explained by pedagogical items, but by mathematics pedagogical items, which are not coded as missings in this category but as "not applicable". The large variance explanation effect for dealing with items concerning recent (mathematics) pedagogical topics is of particular interest, as these comprise, among others, the latest insights related to proficient teaching.

Table 5.7 Effect of experience group membership on the handling of cognitive demands

Categories	Wilks-λ	Part. η2
Teacher noticing facet	.839	.084
Subject-related perspective	.830	.089
Knowledge of mathematics pedagogy	.822	.094
Knowledge of pedagogy	.801	.105
Dealing with heterogeneity	.854	.076
Particular relevance of recent (mathematics) pedagogical topics	.836	.085

Note. Multivariate variance analyses were conducted. $p < .001$ for all MANOVAs.

The pattern that was indicated by mean comparisons (see Table 5.6) is significantly confirmed in the post hoc analyses carried out in the MANOVAs (see Table 5.8 and ESM Table 4). There are mostly highly significant differences ($p < 0.001$) between in-service teachers and MS, but in many categories, there are no significant differences between ECT and ET. Below, we discuss deviations from this pattern and comparisons of consecutive experience groups based on the post hoc tests.

5.2.4.2 Comparisons of the Consecutive Teaching Experience Groups

This section presents the results of comparisons of consecutive experience groups (i.e., MS and ECT as well as ECT and ET). The aspects of decision-making, mathematics pedagogical knowledge, and the particular relevance of recent (mathematics) pedagogical topics will be of particular interest in these analyses.

When contrasting MS directly with ECT (see Table 5.8), significant differences are found for all but one subcategory. These differences vary in effect size from $d = 0.28$ to $d = 1.13$. Mean differences with more than one standard deviation (and corresponding effect sizes) are largest for M_PID ($d = 1.13$), curriculum and planning in mathematics pedagogics ($d = 1.05$), and the particular relevance of recent (mathematics) pedagogical topics ($d = 1.02$). This may indicate particularly large developments in the ability to deal with these demands through the acquisition of teaching experience. No significant difference between the three experience groups was found for students' assessment. Analysis of the four items in this cognitive demand category revealed the items to be comparatively easy and thus non-differentiating tasks.

Since the mean comparisons imply stagnation or regression after some length of teaching experience, comparison of ECT and ET is of particular interest. The analysis (see Table 5.8) revealed some deviations from the pattern of non-significant differences. ECT' significant advantage over ET in dealing with decision-related demands of teacher noticing has already been mentioned ($d = .34$; Bastian et al., 2022). In addition, our new analyses document significant differences in favor of ECT related to the M_PID subject-related perspective ($d = .28$).

Further differences in favor of ECT are observed for cognitive demands in the categories of mathematics pedagogical curriculum and planning ($d = 0.32$), particular focus on inclusive mathematics teaching ($d = 0.28$), students' learning ($d = 0.31$), and particular relevance of recent (mathematics) pedagogical topics ($d = 0.38$). For the latter demand, these differences are significant at the 1% level and show the largest effect size for comparison of the in-service teacher groups. It should be noted that the other three cognitive demands relate, at least in part, to recent mathematics pedagogical or general pedagogical topics. For instance, dealing with new education plans, students' competencies, and basic ideas related to recent educational topics make up a large part of curriculum and planning. In addition, students' learning refers to dealing with heterogeneity

Table 5.8 Post hoc tests for experience group comparisons in select subcategories

Categories	Post Hoc Analyses	Categories	Post Hoc Analyses
Teacher noticing		Knowledge of general pedagogy	
Decision-making	MS < ECT*** MS < ET*** ET < ECT*	Not applicable	MS < ECT*** MS < ET*** ET < ECT*
Subject-related perspective		Students' learning	MS < ECT*** ET < ECT*
M_PID	MS < ECT*** MS < ET*** ET < ECT*	Students' assessment	No significant group differences
Knowledge of mathematics pedagogy		Particular relevance of recent (mathematics) pedagogical topics	
Curriculum & planning	MS < ECT*** MS < ET*** ET < ECT*	Existent	MS < ECT*** MS < ET*** ET < ECT**
Dealing with heterogeneity			
Particularly necessary	MS < ECT*** MS < ET** ET < ECT*		

Note. MS ≙ master's students, ECT ≙ early-career teachers, ET ≙ experienced teachers. Significance of each comparison: * ≙ $p < .05$, ** ≙ $p < .01$, *** ≙ $p < .001$. More statistical information about the post hoc test is given in the ESM Table 4.

as well as adaptive learning methods. Inclusive mathematics teaching can be understood as such a recent pedagogical topic in itself. Therefore, differences in teacher noticing skills between ECT and ET in favor of ECT may be attributed to knowledge and skills related to dealing with recent (mathematics) pedagogical and pedagogical issues. In addition, the difference in the category of students' learning could suggest that ECT have a strong ability to deal with students' individual motivations and cognitions.

5.2.5 Discussion

The present study aimed to investigate which differences exist between the teacher noticing skills—specifically, the ability to deal with the cognitive demands of tasks related to teacher noticing—of three teacher groups with varying levels of teaching experience. These differences were measured by a video-based teacher noticing instrument and evaluated with a developed category system. The category system was conceptually developed based on research-oriented and practice-oriented competence frameworks from relevant literature on teachers' competence domains, cognitive processes, and knowledge facets.

MANOVAs demonstrated that teaching experience has a significant influence for all cognitive demands. Post hoc tests revealed that in-service teachers with teaching experience significantly outperform MS in terms of ability to deal with all cognitive demands, particularly those associated with mathematics pedagogical items and items referring to curriculum and planning. The latter may be explained by the necessity for teaching experience to develop skills in lesson planning (König et al., 2022). Those performance differences between MS and ECT are well documented for German samples in knowledge tests in terms of general pedagogical knowledge, mathematics content knowledge, and mathematics pedagogical content knowledge (e.g., Blömeke et al., 2014; König, 2013; Kleickmann et al., 2013), supporting the current findings for situation-specific skills reported in this study.

Although ECT nominally perform slightly better than ET, these differences are mostly insignificant. However, a detailed analysis of the two groups of in-service teachers demonstrated that there are deviations from this pattern; for some cognitive demands, ECT have a significant lead over ET with small effect sizes. This concerns three areas and groups of cognitive demands. First, significant differences for decision-making might indicate that ECT have stronger skills in the decision-oriented facets of teacher noticing, which lead to higher performance in making professional decisions in the classroom. This unexpected result may imply that ET rely on routinized action patterns and do not question them in situation-specific contexts. However, this could also mark missing expertise of ET in the addressed situations, in which their decision-making was tested, and, thus, indicate a lack of knowledge (Berliner, 2001; Boshuizen et al., 2020; Yang et al., 2021). Second, the results suggest differences in mathematics pedagogical knowledge and its application in the context of teacher noticing in favor of ECT. In other words, mathematical—especially mathematics pedagogical—knowledge is less present in ET. This could imply a decline in professional knowledge among ET, since their initial teacher education, providing formal learning opportunities fostering such declarative-conceptual knowledge explicitly, dates back several years (Liu & Phelps, 2020). It is in line with studies which demonstrate the decline in mathematics content knowledge from ECT to ET using pencil-and-paper tests (Blömeke et al., 2014; Kleickmann et al., 2013). The result emphasizes the essential role of professional knowledge as a prerequisite for the development of expertise in teacher noticing (Gegenfurtner et al., 2020), since ECT, who demonstrate greater expertise in our teacher noticing instrument, particularly excel in task strongly connected to professional knowledge. Third, and of particular interest, significant differences can be observed for the cognitive demands concerning recent mathematics pedagogical and general pedagogical topics, such as inclusive or competence-oriented mathematics education. This raises questions regarding the long-term effects of teacher education, similar to the results of Liu and Phelps (2020), and especially the long-term benefits of respective professional development activities (Goldschmidt & Phelps, 2010).

These results provide further evidence that teacher expertise and teaching experience are by no means equivalent (Stigler & Miller, 2018), as ECT demonstrated greater expertise in teacher noticing than ET with more teaching experience. That could also suggest a shortage of deliberate practice opportunities for ET (Ericsson et al., 1993), especially in recent topics of teaching and learning like competence-orientation. Nonetheless, experience was shown to be a necessary condition for expertise development, as indicated by the large achievement differences between MS without systematic teaching experience and ECT.

The study has limitations, which must be considered. First, the analyzed data are based on convenience samples and were collected cross-sectionally. Thus, statements about generalizations and developments are limited and must be treated with caution. Particularly, some teachers received their teacher education at different times, which might have influenced the development of their teacher noticing and caused different levels of noticing skills. However, descriptive sub-sample characteristics speak for the comparability of the three groups (see Table 5.2). Second, only three groups with different teaching experience were compared, giving a rather rough representation of the stages of teacher education and professional development. Future studies should perform analyses with more differentiated groups or longitudinal designs with several measurement points. Finally, the subsamples were collected over a period from 2011 to 2020. This may have had an unknown influence on the data and, due to the ongoing implementation of teacher noticing fostering in teacher education and professional development (Amador et al., 2021; König et al., 2022; Stahnke et al., 2016), favor groups, whose data were later collected—namely ET and MS.

Despite these limitations, this study aligns with the existing body of research and offers new and detailed results on how groups with different teaching experience deal with cognitive demands when applying their teacher noticing skills, enriching existing findings on expertise development in this area. For example, significant differences between both groups of in-service teachers in terms of recent mathematics pedagogical and general pedagogical topics, and thus modern teaching methods in general, were also found in the Chinese context (Yang et al., 2021). Now, they are replicated for a Western context. This finding may be the result of low participation of ET in appropriate professional development activities (Pedder et al., 2008), low effects of such interventions (Liu & Phelps, 2020), or a lack of expertise in modern mathematics education among ET. Therefore, it is necessary to develop well-designed and, particularly, thoroughly evaluated professional development interventions, especially in areas that had a separating effect on teacher groups in this study (i.e., recent mathematics pedagogical topics and general pedagogical topics like competence-oriented teaching and dealing with heterogeneity).

Supplementary Information: The online version contains supplementary material available at https://doi.org/10.1007/s10763-023-10374-x.

Funding: Open Access funding enabled and organized by Projekt DEAL. TEDS-Follow-Up: German Research Foundation (grant number: KA 797/18–1)

TEDS-Validate: Federal Ministry of Education and Research (grant number: 01PK15006A/B)

TEDS-Validate-Transfer: Federal Ministry of Education and Research (grant number: 01PK19006A/B)

Data Availability: The data are accessible upon request from the authors.

Declarations

Ethical Clearance: Ethical clearance was granted by all participating institutions.

Conflict of Interest: None.

References

Amador, J. M., Bragelman, J., & Superfine, A. C. (2021). Prospective teachers' noticing: A literature review of methodological approaches to support and analyze noticing. *Teaching and Teacher Education*, 99, 1–16. https://doi.org/10.1016/j.tate.2020.103256

Bastian, A., Kaiser, G., Meyer, D., Schwarz, B., & König, J. (2022). Teacher noticing and its growth toward expertise: An expert–novice comparison with pre-service and in-service secondary mathematics teachers. *Educational Studies in Mathematics*, 110(2), 205–232. https://doi.org/10.1007/s10649-021-10128-y

Berliner, D. C. (1994). *Expertise: The wonders of exemplary performance*. In J. N. Mangieri & C. Collins (Eds.), Creating powerful thinking in teachers and students (pp. 141–186). Holt, Rinehart and Winston.

Berliner, D. C. (2001). Learning about and learning from expert teachers. *International Journal of Educational Research*, 35(5), 463–482. https://doi.org/10.1016/S0883-0355(02)00004-6

Blömeke, S., Busse, A., Kaiser, G., König, J., & Suhl, U. (2016). The relation between content-specific and general teacher knowledge and skills. *Teaching and Teacher Education*, 56, 35–46. https://doi.org/10.1016/j.tate.2016.02.003

Blömeke, S., Gustafsson, J. E., & Shavelson, R. J. (2015). Beyond dichotomies: Competence viewed as a continuum. *Zeitschrift Für Psychologie / Journal of Psychology*, 223(1), 3–13. https://doi.org/10.1027/2151-2604/a000194

Blömeke, S., Jentsch, A., Ross, N., Kaiser, G., & König, J. (2022). Opening up the black box: Teacher competence, instructional quality, and students' learning progress. *Learning and Instruction*, 79, 1–11. https://doi.org/10.1016/j.learninstruc.2022.101600

Blömeke, S., Kaiser, G., & Lehmann, R. (Eds.). (2010). *TEDS-M 2008: Professionelle Kompetenz und Lerngelegenheiten angehender Mathematiklehrkräfte für die Sekundarstufe I im internationalen Vergleich* [TEDS-M 2008: Professional competence and opportunities to learn of prospective secondary mathematics teachers in international comparison]. Waxmann.

Blömeke, S., König, J., Suhl, U., Hoth, J., & Döhrmann, M. (2015). Wie situationsbezogen ist die Kompetenz von Lehrkräften? Zur Generalisierbarkeit der Ergebnisse von videobasierten Performanztests [To what extent is teacher competence situation-specific? On the generalizability of the results of video-based performance tests]. *Zeitschrift Für Pädagogik*, 61(3), 310–327.

Blömeke, S., König, J., Busse, A., Suhl, U., Benthien, J., Döhrmann, M., & Kaiser, G. (2014). Von der Lehrerausbildung in den Beruf - Fachbezogenes Wissen als Voraussetzung für Wahrnehmung, Interpretation und Handeln im Unterricht [From teacher education to the profession - subject-related knowledge as a prerequisite for perception, interpretation and decision-making in the classroom]. *Zeitschrift Für Erziehungswissenschaft*, 17(3), 509–542. https://doi.org/10.1007/s11618-014-0564-8

Bond, T. G., Yan, Z., & Heene, M. (2021). *Applying the Rasch model: Fundamental measurement in the human sciences*. Routledge Taylor & Francis Group.

Boshuizen, H. P., Gruber, H., & Strasser, J. (2020). Knowledge restructuring through case processing: The key to generalise expertise development theory across domains? *Educational Research Review*, 29, 1–31. https://doi.org/10.1016/j.edurev.2020.100310

Caspari-Sadeghi, S., & König, J. (2018). On the adequacy of expert teachers: From practical convenience to psychological reality. *International Journal of Higher Education*, 7(5), 1–19. https://doi.org/10.5430/ijhe.v7n5p1

Carter, K., Cushing, K. S., Sabers, D. S., Stein, P., & Berliner, D. C. (1988). Expert-novice differences in perceiving and processing visual classroom information. *Journal ofTeacher Education*, 39(3), 25–31. https://doi.org/10.1177/002248718803900306

Cohen, J. (1960). A coefficient of agreement for nominal scales. *Educational and Psychological Measurement*, 20(1), 37–46. https://doi.org/10.1177/001316446002000104

Cohen, J. (1988). *Statistical power analysis for the behavioral sciences* (2nd ed.). Lawrence Erlbaum Associates.

Copur-Gencturk, Y., & Tolar, T. (2022). Mathematics teaching expertise: A study of the dimensionality of content knowledge, pedagogical content knowledge, and content-specific noticing skills. *Teaching and Teacher Education*, 114, 103696. https://doi.org/10.1016/j.tate.2022.103696

Dietiker, L., Males, L. M., Amador, J. M., & Earnest, D. (2018). Curricular noticing: A framework to describe teachers' interactions with curriculum materials. *Journal for Research in Mathematics Education*, 49(5), 521–532. https://doi.org/10.5951/jresemath educ.49.5.0521

Dindyal, J., Schack, E. O., Choy, B. H., & Sherin, M. G. (2021). Exploring the terrains of mathematics teacher noticing. *ZDM Mathematics Education*, 53(1), 1–16. https://doi.org/10.1007/s11858-021-01249-y

Dreher, A., & Kuntze, S. (2015). Teachers' professional knowledge and noticing: The case of multiple representations in the mathematics classroom. *Educational Studies in Mathematics*, 88(1), 89–114. https://doi.org/10.1007/s10649-014-9577-8

Ericsson, K. A., Krampe, R. T., & Tesch-Römer, C. (1993). The role of deliberate practice in the acquisition of expert performance. *Psychological Review*, 100(3), 363–406. https://doi.org/10.1037/0033-295X.100.3.363

Gegenfurtner, A. (2020). *Professional vision and visual expertise*. Universität Regensburg.

Gegenfurtner, A., Lewalter, D., Lehtinen, E., Schmidt, M., & Gruber, H. (2020). Teacher expertise and professional vision: Examining knowledge-based reasoning of pre-service teachers, in-service teachers, and school principals. *Frontiers in Education*, 5(59). https://doi.org/10.3389/feduc.2020.00059

Gold, B., & Holodynski, M. (2017). Using digital video to measure the professional vision of elementary classroom management: Test validation and methodological challenges. *Computers & Education*, 107, 13–30. https://doi.org/10.1016/j.compedu.2016.12.012

Goldschmidt, P., & Phelps, G. (2010). Does teacher professional development affect content and pedagogical knowledge: How much and for how long? *Economics of Education Review*, 29, 432–439. https://doi.org/10.1016/j.econedurev.2009.10.002

Goodwin, C. (1994). Professional vision. American Anthropologist, 96(3), 606–633.

Gruber, H., Scheumann, M., & Krauss, S. (2019). Problemlösen und Expertiseerwerb [Problem solving and acquisition of expertise]. In D. Urhahne, M. Dresel, & F. Fischer (Eds.), *Psychologie für den Lehrberuf* (pp. 53–65). Springer.

Hoth, J., Schwarz, B., Kaiser, G., Busse, A., König, J., & Blömeke, S. (2016). Uncovering predictors of disagreement: ensuring the quality of expert ratings. *ZDM Mathematics Education*, 48(1-2), 83–95. https://doi.org/10.1007/s11858-016-0758-z

Jacobs, V. R., Lamb, L. L. C., & Philipp, R. A. (2010). Professional noticing of children's mathematical thinking. *Journal for Research in Mathematics Education*, 41(2), 169–202. www.jstor.org/stable/20720130

Kaiser, G., Blömeke, S., König, J., Busse, A., Döhrmann, M., & Hoth, J. (2017). Professional competencies of (prospective) mathematics teachers—cognitive versus situated approaches. *Educational Studies in Mathematics*, 94(2), 1–22. https://doi.org/10.1007/s10649-016-9713-8

Kaiser, G., Busse, A., Hoth, J., König, J., & Blömeke, S. (2015). About the complexities of video-based assessments: Theoretical and methodological approaches to overcoming shortcomings of research on teachers' competence. *International Journal of Science and Mathematics Education*, 13(2), 369–387. https://doi.org/10.1007/s10763-015-9616-7

Kleickmann, T., Richter, D., Kunter, M., Elsner, J., Besser, M., Krauss, S., & Baumert, J. (2013). Teachers' content knowledge and pedagogical content knowledge. *Journal of Teacher Education*, 64(1), 90–106. https://doi.org/10.1177/0022487112460398

König, J. (2013). First comes the theory, then the practice? On the acquisition of general pedagogical knowledge during initial teacher education. *International Journal of Science and Mathematics Education*, 11(4), 999–1028. https://doi.org/10.1007/s10763-013-9420-1

König, J., & Kramer, C. (2016). Teacher professional knowledge and classroom management: On the relation of general pedagogical knowledge (GPK) and classroom management expertise (CME). *ZDM Mathematics Education*, 48(1), 139–151.

König, J., Cammann, F., Bremerich-Vos, A., & Buchholtz, C. (2022). Unterrichtsplanungskompetenz von (angehenden) Deutschlehrkräften der Sekundarstufe: Testkonstruktion und Validierung [Lesson planning competence of (prospective) secondary language teachers: Test construction and validation]. *Zeitschrift Für Erziehungswissenschaft*, 25(4), 869–894. https://doi.org/10.1007/s11618-022-01113-z

König, J., Gerhard, K., Melzer, C., Rühl, A.-M., Zenner, J., & Kaspar, K. (2017). Erfassung von pädagogischem Wissen für inklusiven Unterricht bei angehenden Lehrkräften: Testkonstruktion und Validierung [Assessing pedagogical content knowledge for inclusive education of prospective teachers: Test construction and validation]. *Unterrichtswissenschaft*, 45(4), 223–242.

König, J., Santagata, R., Scheiner, T., Adleff, A.-K., Yang, X., & Kaiser, G. (2022). Teacher noticing: A systematic literature review of conceptualizations, research designs, and findings on learning to notice. *Educational Research Review*, 36. https://doi.org/10.1016/j.edurev.2022.100453

Kramer, C., König, J., Strauß, S., & Kaspar, K. (2020). Classroom videos or transcripts? A quasi-experimental study to assess the effects of media-based learning on pre-service teachers' situation-specific skills of classroom management. *International Journal of Educational Research*, 103, 101624. https://doi.org/10.1016/j.ijer.2020.101624

Krauss, S. (2020). Expertise-Paradigma in der Lehrerinnen- und Lehrerbildung [The expertise paradigm in teacher education]. In C. Cramer, J. König, M. Rothland, & S. Blömeke (Eds.), Handbuch Lehrerinnen- und Lehrerbildung (pp. 154–162). UTB.

Kuckartz, U. (2014). *Qualitative text analysis: A guide to methods, practice & using software*. SAGE. https://doi.org/10.4135/9781446288719

Kunter, M., Baumert, J., Blum, W., Klusmann, U., Krauss, S., & Neubrand, M. (Eds.). (2011). *Professionelle Kompetenz von Lehrkräften: Ergebnisse des Forschungsprogramms*

COACTIV [Professional competence of teachers: results from the COACTIV research program]. Waxmann.
Lachner, A., Jarodzka, H., & Nückles, M. (2016). What makes an expert teacher? Investigating teachers' professional vision and discourse abilities. *Instructional Science*, 44(3), 197–203. https:// doi.org/https://doi.org/10.1007/s11251-016-9376-y
Landis, J. R., & Koch, G. G. (1977). The measurement of observer agreement for categorical data. *Biometrics*, 33(1), 159. https://doi.org/10.2307/2529310
Liu, S., & Phelps, G. (2020). Does teacher learning last? Understanding how much teachers retain their knowledge after professional development. *Journal of Teacher Education*, 71(5), 537–550. https://doi.org/10.1177/0022487119886290
Louie, N., Adiredja, A. P., & Jessup, N. (2021). Teacher noticing from a sociopolitical perspective: The FAIR framework for anti-deficit noticing. *ZDM Mathematics Education*, 53(1), 95–107. https://doi.org/10.1007/s11858-021-01229-2
Martin, M., Farrell, M., Seidel, T., Rieß, W., Könings, K. D., van Merriënboer, J. J. G., & Renkl, A. (2022). Focused self-explanation prompts and segmenting foster pre-service teachers' professional vision - but only during training! *International Journal of Educational Technology in Higher Education*, 19(1), 1–29. https://doi.org/10.1186/s41239-022-00331-z
Mason, J. (2002). *Researching your own practice: The discipline of noticing*. Routledge.
McDonald, S. P. (2016). The transparent and the invisible in professional pedagogical vision for science teaching. *School Science and Mathematics*, 116(2), 95–103. https://doi.org/10.1111/ssm.12156
Meschede, N., Fiebranz, A., Möller, K., & Steffensky, M. (2017). Teachers' professional vision, pedagogical content knowledge and beliefs: On its relation and differences between pre-service and in-service teachers. *Teaching and Teacher Education*, 66, 158–170. https://doi.org/10.1016/j.tate.2017.04.010
Metsäpelto, R.-L., Poikkeus, A.-M., Heikkilä, M., Husu, J., Laine, A., Lappalainen, K., Lähteenmäki, M., Mikkilä-Erdmann, M., Warinowski, A., Iiskala, T., Hangelin, S., Harmoinen, S., Holmström, A., Kyrö-Ämmälä, O., Lehesvuori, S., Mankki, V., & Suvilehto, P. (2021). A multidimensional adapted process model of teaching. *Educational Assessment, Evaluation and Accountability*, 1–30. https://doi.org/10.1007/s11092-021-09373-9
Palmer, D. J., Stough, L. M., Burdenski, T. K., Jr., & Gonzales, M. (2005). Identifying teacher expertise: An examination of researchers' decision making. *Educational Psychologist*, 40(1), 13–25. https://doi.org/10.1207/S15326985EP4001_2
Pedder, D., Storey, A., Opfer, V. D., & Wolfenden, F. (2008). *Schools and continuing professional development (CPD) in England - State of the Nation research project*. Cambridge University and the Open University.
Piwowar, V., Barth, V. L., Ophardt, D., & Thiel, F. (2018). Evidence-based scripted videos on handling student misbehavior: The development and evaluation of video cases for teacher education. *Professional Development in Education*, 44(3), 369–384. https://doi.org/10.1080/19415257.2017.1316299
Resnick, L. B. (1975). *Task analysis in instructional design: Some cases from mathematics*. Pittsburgh University. https://eric.ed.gov/?id=ED115486
Santagata, R., & Yeh, C. (2016). The role of perception, interpretation, and decision making in the development of beginning teachers' competence. *ZDM Mathematics Education*, 48(1–2), 153–165.

Santagata, R., König, J., Scheiner, T., Nguyen, H., Adleff, A.-K., Yang, X., & Kaiser, G. (2021). Mathematics teacher learning to notice: A systematic review of studies of video-supported teacher education. *ZDM Mathematics Education*, 53, 119–134. https://doi.org/10.1007/s11858-020-01216-z

Schack, E. O., Fisher, M. H., & Wilhelm, J. A. (Eds.). (2017). *Research in Mathematics Education. Teacher noticing: Bridging and broadening perspectives, contexts, and frameworks.* Springer International Publishing. https://doi.org/10.1007/978-3-319-46753-5

Seidel, T., & Stürmer, K. (2014). Modeling and measuring the structure of professional vision in pre-service teachers. *American Educational Research Journal*, 51(4), 739–771. https://doi.org/10.3102/0002831214531321

Seidel, T., Schnitzler, K., Kosel, C., Stürmer, K., & Holzberger, D. (2021). Student characteristics in the eyes of teachers: Differences between novice and expert teachers in judgment accuracy, observed behavioral cues, and gaze. *Educational Psychology Review*, 33(1), 69–89. https://doi.org/10.1007/s10648-020-09532-2

Sherin, M. G., Jacobs, V. R., & Philipp, R. A. (2011). Situating the study of teacher noticing. In M.G. Sherin, V. R. Jacobs, & R. A. Philipp (Eds.), *Studies in mathematical thinking and learning. Mathematics teacher noticing* (pp. 3–13). Routledge.

Stahnke, R., & Blömeke, S. (2021). Novice and expert teachers' situation-specific skills regarding classroom management: What do they perceive, interpret and suggest? *Teaching and Teacher Education*, 98, 103243. https://doi.org/10.1016/j.tate.2020.103243

Stahnke, R., Schueler, S., & Roesken-Winter, B. (2016). Teachers' perception, interpretation, and decision-making: A systematic review of empirical mathematics education research. *ZDM Mathematics Education*, 48(1– 2), 1–27. https://doi.org/10.1007/s11858-016-0775-y

Stigler, J. W., & Miller, K. F. (2018). Expertise and expert performance in teaching. In K. A. Ericsson, R. R. Hoffman, A. Kozbelt, & M. A. Williams (Eds.), *The Cambridge handbook of expertise and expert performance* (pp. 431–452). Cambridge University Press. https://doi.org/10.1017/9781316480748.024

Turner, R., Blum, W., & Niss, M. (2015). Using competencies to explain mathematical item demand: A work in progress. In K. Stacey & R. Turner (Eds.), *Education. Assessing mathematical literacy: The PISA experience* (pp. 85–115). Springer Spektrum. https://doi.org/10.1007/978-3-319-10121-7_4

van Es, E. A. (2011). A framework for learning to notice student thinking. In M. G. Sherin, V. R. Jacobs, & R. A. Philipp (Eds.), *Studies in mathematical thinking and learning. Mathematics teacher noticing* (pp. 134–151). Routledge.

van Es, E. A., & Sherin, M. G. (2002). Learning to notice: Scaffolding new teachers' interpretations of classroom interactions. *Journal of Technology and Teacher Education*, 10, 571–596.

van Es, E. A., & Sherin, M. G. (2021). Expanding on prior conceptualizations of teacher noticing. *ZDM Mathematics Education* 53, 17-27. https://doi.org/10.1007/s11858-020-01211-4

Walkoe, J. (2015). Exploring teacher noticing of student algebraic thinking in a video club. *Journal of Mathematics Teacher Education*, 18(6), 523–550. https://doi.org/10.1007/s10857-014-9289-0

Weinert, F. E. (2001). Concept of Competence: A conceptual clarification. In D. Rychen & L. Salganik (Eds.), *Defining and selecting key competencies* (pp. 45–65). Hogrefe & Huber.

Weyers, J., König, J., Santagata, R., Scheiner, T., & Kaiser, G. (2023). Measuring teacher noticing: A scoping review of standardized instruments. *Teaching and Teacher Education, 122*. https://doi.org/10.1016/j.tate.2022.103970

Wolff, C. E., Jarodzka, H., van den Bogert, N., & Boshuizen, H. P. A. (2016). Teacher vision: Expert and novice teachers' perception of problematic classroom management scenes. *Instructional Science*, 44(3), 243–265. https://doi.org/10.1007/s11251-016-9367-z

Yang, X., Kaiser, G., König, J., & Blömeke, S. (2018). Measuring Chinese teacher professional competence: adapting and validating a German framework in China. *Journal of Curriculum Studies*, 50(5), 638–653. https://doi.org/10.1080/00220272.2018.1502810

Yang, X., König, J., & Kaiser, G. (2021). Growth of professional noticing of mathematics teachers: A comparative study of Chinese teachers noticing with different teaching experience. *ZDM Mathematics Education*, 53, 29–42. https://doi.org/10.1007/s11858-020-01217-y

Publication III

6

In 2024, the paper "Effects of teaching internships on preservice teachers' teacher noticing in secondary mathematics education" was published in the journal Frontiers in Education as the third publication for this thesis. The research was conducted as part of the TEDS-V-T study, which focused on the transferability of the TEDS teacher noticing instrument developed for in-service teachers to measure the teacher noticing skills of pre-service teachers. In a first publication, Weyers, König, Rott, et al. (2023) had already investigated general transferability and linked university opportunities to learn with teacher noticing skills. As a follow-up, in Publication III of this thesis, we focused on the pre-posttest applicability and examined the influences of the teaching internship at the end of university teacher education on teacher noticing as well as the impacts of characteristics of the practice experience, namely opportunities to learn, on situation-specific skills in a pretest-posttest design. In doing so, the paper addresses the lack of studies on the effects of teaching internships using standardized measures and longitudinal studies in the field of teacher noticing in general. In the expanded impact model of teacher education, the paper was located between area (1) "Teacher education and professional development" and area (2) "Teacher competence," with important reference to situation-specific skills, learning opportunities, and teacher internships (see Fig. 6.1).

In a pretest-posttest design, we investigated the effects of teaching internships and related opportunities to learn, such as conducting lessons or reflecting on teaching practice, on the three facets of teacher noticing, perception, interpretation, and decision-making. In addition, cross-lagged effects of these facets were examined to include reciprocal influences of the facets on each other and to facilitate insights into the development of teacher noticing and how its three facets can

A. Bastian, *Teacher Noticing of Pre-service and In-service Secondary Mathematics Teachers*, Perspektiven der Mathematikdidaktik,
https://doi.org/10.1007/978-3-658-45619-1_6

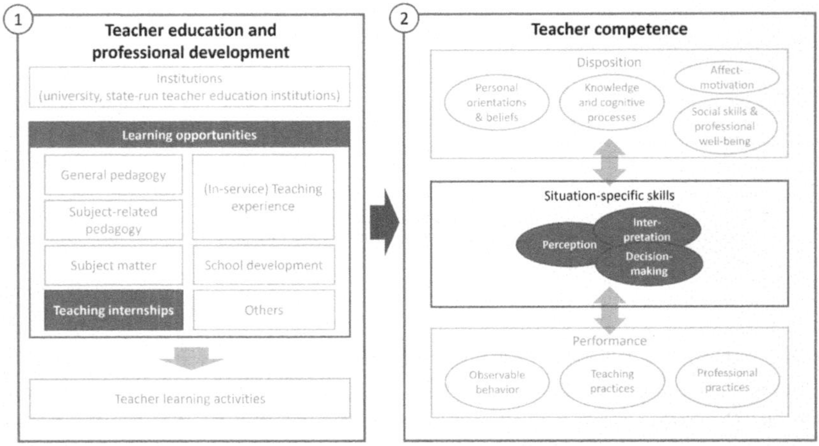

Note. **Areas examined by the research presented in Publication III are marked in blue.**

Fig. 6.1 Excerpt of the Expanded Impact Model to Illustrate the Contribution of Publication III

predict this development. This study addressed the research questions of whether and to what extent teacher noticing changes over the course of a teaching internship and how teaching internship process variables influence changes in teacher noticing skills. We analyzed a sample of 175 master's students from six German universities using cross-lagged panel analyses, multiple regression, and paired sample t-tests. Data scaling was conducted using IRT approaches.

The results indicated a significant improvement in all three facets of teacher noticing over the course of the internship with small effect sizes, and interpretation was a key facet in this development, having an autoregressive impact as well as influencing the development of perception and decision-making. Only some opportunities to learn within the teacher internship showed a significant impact on teacher noticing skills. For instance, connecting theory and practice and reflecting on practice seemed to foster teacher noticing skills, while the sole process of teaching had no effects on interpretation or decision-making, and even had a negative effect on perception. Overall, the study demonstrated the potential of teaching internships for the development of master's students' teacher noticing and highlighted areas for improvement.

6.1 Individual Contribution Statement

In the course of the TEDS-V-T study, Gabriele Kaiser, Johannes König, Jonas Weyers, Hans-Stefan Siller and I collaborated on the research reported in Publication III. Gabriele Kaiser and Johannes König planned the publication as a key contribution to the project TEDS-V-T. I specified the publication idea and planned the necessary empirical analyses in close collaboration with Jonas Weyers. Data collection and cleaning was mainly done by Dennis Sommer. I scaled the data using Rasch models and conducted all statistical analyses, most notably cross-lagged panel analyses and the multiple regressions, with the help of Jonas Weyers. I wrote the complete draft of the paper. Subsequently, I revised the paper based on feedback from all co-authors and prepared the paper for submission. Gabriele Kaiser submitted the manuscript to the journal. All revisions were made by me in close collaboration with Gabriele Kaiser. I also presented the research at the conference of the Gesellschaft für Empirische Bildungsforschung [Society for Empirical Educational Research] in 2023.

6.2 Publication Imprint

Effects of teaching internships on preservice teachers' noticing in secondary mathematics education

Anton Bastian[1] (ORCID: 0000-0002-1177-6336), Johannes König[2] (ORCID: 0000-0003-3374-9408), Jonas Weyers[2] (ORCID: 0000-0003-1804-4434), Hans-Stefan Siller[3] (ORCID: 0000-0003-1597-7108), Gabriele Kaiser[1,4*] (ORCID: 0000-0002-6239-0169)

[1]Department of Mathematics Education, Faculty of Education, University of Hamburg, Hamburg, Germany
[2]Department of Educational and Social Studies, Faculty of Human Sciences, University of Cologne, Cologne, Germany

[1] Bastian, A., König, J., Weyers, J., Siller, H.-S., & Kaiser, G. (2024). Effects of teaching internships on preservice teachers' noticing in secondary mathematics education. *Frontiers in Education*. https://doi.org/10.3389/feduc.2024.1360315

[3]Institute of Mathematics, Faculty of Mathematics and Computer Science, University of Würzburg, Würzburg, Germany
[4]Faculty of Education and Arts, Nord University, Bodø, Norway
* **Correspondence:** Gabriele Kaiser, gabriele.kaiser@uni-hamburg.de

Keywords: initial teacher education, teacher noticing, teaching internships, teaching practice, teaching practicum, mathematics secondary education, preservice teachers

Abstract
In this study, we examined the effects of teaching internships and related opportunities to learn, such as conducting lessons or reflecting on teaching practice, on the three facets of teacher noticing, perception, interpretation, and decision-making. Cross-lagged effects of these facets were examined to include reciprocal influences of the facets on each other and to facilitate insights into the development of teacher noticing and how its three facets can predict this development. In detail, this study addressed the research questions of whether and to what extent teacher noticing changes over the course of a teaching internship and how teaching internship process variables influence changes in teacher noticing skills. Based on a sample of 175 preservice teachers from six German universities, we studied professional noticing using a video-based pre- and posttest approach. The results indicated a significant improvement in all three facets of teacher noticing over the course of the internship with small effect sizes, and interpretation was a key facet in this development, having an autoregressive impact as well as influencing the development of perception and decision-making. Only some opportunities to learn within the teacher internship showed a significant impact on teacher noticing skills. For instance, connecting theory and practice and reflecting on practice seemed to foster teacher noticing skills, while the sole process of teaching had no effects on interpretation or decision-making, and even had a negative effect on perception. Overall, the study demonstrated the potential of teaching internships for the development of preservice teachers' noticing skills and highlighted areas for improvement.

6.2.1 Introduction

An important goal of the university-based phase of teacher education is to provide learning opportunities for preservice teachers (PSTs) so that they can acquire the knowledge and skills necessary for their professional school practice and, in particular, its core, sound teaching skill (Potari & Chapman, 2020). However, research suggests that PSTs often struggle to recognize their university learning as relevant to their future professional practice as teachers, and they thus demand more school practice and practice-oriented courses to alleviate the tension between theory and practice (Hascher, 2014; Ulrich et al., 2020). However, it may occur as risk that PSTs merely adopt established school practices without reflecting on them (Chitpin et al., 2008; Özgün-Koca & İlhan Şen, 2006; Ulrich et al., 2020). This risk can be mitigated by integrating practical activities into university studies via teaching internships[2] and a strong focus on theory–practice linkage, that is, the establishment of links between theory and practice and the contextualization of theory in practice (König & Blömeke, 2012; Scholten & Orschulik, 2022).

Phases of school practice based on this premise are characterized by university educators providing institutionalized support for PSTs or specialized state-run teacher education institutions offering theory, knowledge, and guidance to accompany practical experiences of teaching, although the extent of this support varies (Lawson et al., 2015; Terhart, 2021; Ulrich et al., 2020). Classroom teaching as a core activity of this school practice involves, among others, perceiving classroom activities, attending to meaningful events, dealing with complex and challenging situations, and further processing and incorporating them into teaching practice (Sherin et al., 2011; van Es & Sherin, 2002). These skills are often conceptualized as teacher noticing, also referred to as professional vision[3] (König, Santagata, et al., 2022) and constitute a central, situation-specific facet of teachers' competence that is more proximal to students' learning processes and teachers' performance in classroom situations than knowledge (Blömeke et al., 2022). To learn effective teaching and progress in their expertise development as teachers, PSTs must develop teacher noticing skills (Bastian et al., 2022; Berliner, 1988; Fernandez & Choy, 2019). However, the structure and characteristics of teachers' noticing remain debatable among researchers, as different theoretical perspectives and research traditions have shaped the academic discourse on conceptualizations and the development of teacher noticing (Amador et al., 2021;

[2] In this study, we use the term *teaching internship* to describe long-term practical activities integrated in teacher education at universities.

[3] In the current discourse, the terms *teacher noticing* and *professional vision* are employed as comparable and overlapping, situation-specific constructs (Weyers, König, Santagata, et al., 2023). In the current study, we use the more common term (teacher noticing; König, Santagata, et al., 2022) to describe this construct and our own conceptualization of it. However, since the characteristics of the facets delineated in the research on noticing and professional vision vary, we employ the authors' terminology when referring to specific studies.

Dindyal et al., 2021; Fernandez & Choy, 2019; König, Santagata, et al., 2022). A cognitive–psychological perspective, which provided the theoretical foundation for this study, has been identified as the most influential perspective in the current discourse and is often used in standardized research on teacher noticing (König, Santagata, et al., 2022; Weyers, König, Santagata, et al., 2023). This perspective views teachers' noticing as a set of cognitive processes that occur within individual teachers during classroom instruction (Jacobs et al., 2010; Seidel & Stürmer, 2014; van Es & Sherin, 2002).

In contrast, a sociocultural perspective emphasizes the social construction of professional vision (Goodwin, 1994), a discipline-specific approach introduced by Mason (2002) focuses on fostering individual teacher's noticing, and an expertise-oriented perspective rooted in the expertise paradigm (Berliner, 1988; Stigler & Miller, 2018) investigates differences between novices' and experts' perceptions and interpretations, and the development of teachers' noticing toward expertise. Emerging approaches place greater emphasis on the sociopolitical contexts and reciprocity of teacher noticing. For example, Louie et al. (2021) examined framing as an essential part of noticing in their FAIR framework, Scheiner (2021) highlighted the active, exploratory role of teachers in noticing and the reciprocal nature of perceivers and their environment with the embodied ecological approach, and Dominguez (2019) discussed noticing as a reciprocal process between teachers and students.

Despite these different perspectives on the construct, PST's noticing skills are generally expected to develop through school practice as part of teaching internships in initial teacher education and to indicate change over the course of teaching internships, since these skills are deemed to mediate between PSTs' dispositions and performance, and thus also to link theory to practice (Blömeke, Gustafsson, & Shavelson, 2015; Mertens & Gräsel, 2018).

However, despite the fact that teaching internships are the most expensive and organizationally challenging components of initial teacher education, empirical evidence on the effects of teaching internships in general and particularly on teacher noticing is somewhat limited and often based on datasets drawn from single universities. Furthermore, most relevant studies have only examined PSTs' self-assessments of their competencies (König & Rothland, 2018). Thus, to date, findings on the impact of teaching internships on PSTs' development have rarely employed standardized, longitudinal measures of competencies going beyond the own institution (Lawson et al., 2015; Ulrich et al., 2020). Moreover, the development of teacher noticing skills and, in particular, the effects of process variables, such as opportunities to learn during internships, on teacher noticing have scarcely been investigated (Ulrich et al., 2020). In general, data from longitudinal pretest–posttest studies on the development of teachers' noticing skills are limited (König, Santagata, et al., 2022; Weyers, König, Santagata, et al., 2023), although they are needed to explore connections among the development of the facets of teachers' noticing (perception, interpretation, and decision-making) and how they support each other over time, particularly over the course of teaching internships (Superfine et al., 2017; Thomas et al., 2021).

In this study, we aimed to address these research gaps concerning teaching internship by analyzing German secondary PSTs' pre- and posttest results based on an established and standardized video-based measurement instrument for teacher noticing from a general pedagogical and mathematics pedagogical perspective, administered before and after their teaching internships, and relating them to the characteristics of the individual internship experiences of the PSTs. Specifically, we examined the reciprocal effects of the facets of teacher noticing (research question 1), changes in PSTs' noticing skills (research question 2), and the influences of opportunities to learn on these changes (research question 3) over the course of the teaching internship at six German universities. In doing so, we provide new empirical insight into school experiences in initial teacher education that are an essential part of teacher education worldwide (Cabaroğlu & Öz, 2023; E. Cohen et al., 2013; Lawson et al., 2015), and empirical evidence on the extent to which teaching internships can promote teacher noticing skills as a core component of teacher competence.

6.2.2 State of the Art, Theoretical Framework, and Research Questions

In this section, we present the current state of the art research on teacher noticing and describe our own conceptualization. We then summarize relevant research on the effects of teaching internships in general and on teacher noticing in particular. Finally, we describe our research questions.

6.2.2.1 Teacher Noticing

Major research themes in teacher noticing include, in particular, mathematics and mathematics pedagogical topics such as students' mathematical thinking (Jacobs et al., 2010) or ways of dealing with representations (Dreher & Kuntze, 2015), and general pedagogical topics, such as classroom management (Gold & Holodynski, 2017; Weber et al., 2018). Recent studies exploring teachers' noticing have considered dealing with heterogeneity (Keppens et al., 2021) and sociopolitical dimensions, such as ethnicity or socioeconomic background (Louie et al., 2021; Shah & Coles, 2020), as well as strength-oriented noticing (Scheiner, 2023).

Based on a cognitive–psychological approach, teacher noticing is widely understood as paying attention to or perceiving classroom events, interpreting these events, and—in some conceptualizations—employing decision-making processes to act/react based on the interpretations (Dindyal et al., 2021; König, Santagata, et al., 2022; Sherin et al., 2011). For example, Jacobs et al. (2010) focused in their seminal work on children's mathematical understanding, distinguishing three facets of teachers' noticing: attending to children's mathematical strategies, interpreting their understanding, and deciding how to respond. Although the facets of teachers' noticing support each other and may give the impression that noticing in the classroom is a linear or even chronological process (van Es, 2011), from a theoretical perspective, they are understood as deeply "interrelated and cyclical" (Sherin et al., 2011, p. 5), operating through complex interactions (Kaiser et al., 2015; Santagata & Yeh, 2016; Thomas et al., 2021). This circular modeling of in-the-moment teachers' noticing is often expressed as perception $\leftrightarrow$ interpretation $\leftrightarrow$ decision-making (Santagata & Yeh, 2016; Thomas et al., 2021).

However, little empirical attention has been paid to these relationships and their nature, despite knowledge of them being crucial for comprehending teachers' noticing and its effects in the classroom (Thomas et al., 2021). Some studies have claimed that perception and interpretation, and interpretation and decision-making, are more closely related than perception and decision-making, providing some insight into the internal structure of the construct teacher noticing (Bastian et al., 2022; Thomas et al., 2021), but less is known about the development of the facets' relationships over time and how they influence and change each other. This deficiency may be explained by the lack of quantitative pretest–posttest studies on teacher noticing (König, Santagata, et al., 2022).

One exception is a study by Jong et al. (2021), who examined changes in PSTs' noticing skills over the course of one semester. They reported a significant effect of perception at Time 1 (T1) on perception at Time 2 (T2, $\beta = .27$) and similar results for interpretation ($\beta = .35$), but no such association between decision-making at T1 and T2. In addition, only one cross-lagged path was significant (the one from interpretation at T1 to perception at T2 [$\beta = .16$]), indicating an effect of interpretation on perceptual skills. These results were consistent with a study by Superfine et al. (2017), which demonstrated that it may be better to foster interpretation skills before perceptual skills. In our study, we further analyzed the cross-lagged, reciprocal effects of the three facets of teacher noticing to provide a clearer picture of their relationships.

6.2.2.2 Conceptualization of teacher noticing in the Teacher Education and Development Study in Mathematics (TEDS-M) research program

Competencecan be conceptualized as a set of cognitive skills and abilities needed to successfully cope with the demands of professional situations, including the motivational, volitional, and social willingness to apply them (Weinert, 2001). Following this conceptualization, competence research at first particularly focused on cognitive dispositions of teacher such as their professional knowledge (Kunter et al., 2013). To consider the situation-specificity of competence, conceptualizations have been further developed in recent years based on complementary components, such as teacher noticing (Kaiser et al., 2017; Metsäpelto et al., 2021). For example, Blömeke, Gustafsson, and Shavelson (2015) conceptualized competence as a continuum between dispositions (e.g., knowledge) and actual classroom performance, and they included perceiving, interpreting, and decision-making as mediating skills to emphasize the situation-specific aspects of competence. These skills, which we describe as teacher noticing, contextualize knowledge and affects and link them to teachers' actual performance. Thus, they also connect theory learned in initial teacher education to practice in the classroom.

The TEDS-M research program includes situation-specific skills in the measurement of teachers' professional competencies by incorporating teacher noticing as an integral part in the competence framework since the TEDS-Follow-Up (TEDS-FU) study (Kaiser et al., 2017). Based on the model by Blömeke, Gustafsson, and Shavelson (2015), we conceptualize teacher noticing as consisting of three facets: (1) perception of specific classroom events, (2) interpretation of the perceived events, and (3) decision-making, that is, the preparation of responses to student actions or alternative instructional strategies (Kaiser et al., 2015). This can be classified as an analytical and cognitive–psychological conceptualization of teacher noticing (König, Santagata, et al., 2022). A cognitive-psychological approach was chosen to model and measure the cognitive processes involved in teacher noticing and to compare the skills of groups of teachers as well as the development of individual teachers over time (König, Santagata, et al., 2022). The approach allows for a standardized and feasible operationalization of the noticing facets and is thus commonly applied in the psychometric measurement of teacher noticing (Weyers, König, Santagata, et al., 2023).

Based on studies from the expertise research (Berliner, 1988; Stigler & Miller, 2018), the first facet of teacher noticing in our conceptualization involves perceptual processes with no or only minimal interpretation (Bastian et al., 2022), encompassing the observation of clearly discernable incidents that happen in classroom. The second facet, interpretation, involves the analysis of observed events based on an individual's knowledge, experiences, and beliefs, excluding the consideration of instructional responses to these events. This, in turn, is part of the third facet of teacher noticing that includes the development of possible continuations of the lesson, responses to student behavior, and alternative approaches to observed teacher actions (Yang et al., 2021). At the content level, our conceptualization, used in this study, does not focus only on one topic, such as children's mathematical understanding, but encompasses a broad field of classroom situations and features that are relevant to high-quality mathematics education from general and mathematics pedagogical perspectives, such as perceiving effective classroom management, analyzing the use of different representations, interpreting students' mathematical thinking, or making decisions about teaching and learning processes (Kaiser et al., 2015; Yang et al., 2021).

6.2.2.3 Effects of Teaching Internships during University Teacher Education

Over the past 20 years, a growing body of research has considered teaching internships and their impact on all stakeholders, particularly PSTs (see reviews by Lawson et al., 2015; Ulrich et al., 2020). Early studies revealed the crucial role of teaching internships in the self-perceived personal and professional development of PSTs (e.g., Caires & Almeida, 2005), and more recent literature has demonstrated the positive impact of teaching internships on the beliefs, self-efficacy, and motivation of PSTs (García-Lázaro et al., 2022; Ng et al., 2010; Seifert & Schaper, 2018). Mentors foster the development of their mentees and play a crucial role in how PSTs use and perceive their teaching internships (Festner et al., 2020; García-Lázaro et al., 2022; Hudson & Millwater, 2008; König, Darge, Kramer, et al., 2018). In addition, research has suggested that professional competencies, which, in this context, are often distinguished as teaching, educating, assessing, and innovating, are expanded during teaching internships (Gröschner et al., 2013; Seifert et al., 2018).

However, these studies have usually been based solely on PSTs' self-assessments of their own competencies and have not applied objective tests to measure PSTs' professional competencies (Ulrich et al., 2020). Consequently, evidence of competence gains is limited, although there are some exceptions. Recent studies within the *Learning to Practice* project have used knowledge tests to assess PSTs' general pedagogical knowledge, have combined and linked these with self-assessment measures, and demonstrated improvements in complex decision-related knowledge among student teachers from three different universities. Reflection and theory–practice links were shown to facilitate this improvement (König, Darge, Klemenz, & Seifert, 2018). Furthermore, the incorporation of reflection activities and the establishment of links between theoretical concepts and school practice have the potential to enhance knowledge development, as shown by Schlag and Glock (2019). They conducted an analysis of classroom management knowledge using standardized tests and highlighted the significance of practice activities in the acquisition of classroom management knowledge. In summary, although teaching internships have been examined in several studies, research gaps remain regarding the investigation of teachers' competence development using standardized measures (Ulrich et al., 2020).

To date, situation-specific facets of competence, such as teacher noticing, have scarcely been investigated in the teaching internship context (Ulrich et al., 2020). However, in a more general context, comparisons between teachers with different durations of experience have emphasized the critical role of teaching experiences in the development of teachers' noticing (e.g., Bastian et al., 2022; Gold & Holodynski, 2017; Jacobs et al., 2010; Yang et al., 2021). In these studies, PSTs performed worse than in-service teachers on standardized tests of teachers' noticing (Bastian et al., 2022). Thus, practical experiences in schools (e.g., teaching internships) should facilitate the development of PSTs' teacher noticing. This hypothesis was confirmed in a study by Stürmer et al. (2013), who investigated a five-month teaching internship at a German university, focusing on perceptual and interpretive facets of teachers' general pedagogical noticing conceptualized as professional vision. They demonstrated increases ($\eta^2 = .09$, corresponding to an effect size of $d = .63$ [Lenhard & Lenhard, 2016–2022]) in both facets using the video-based test *Observer* (Seidel & Stürmer, 2014).

These findings were replicated by Mertens and Gräsel (2018) for a holistic understanding of professional vision, which included perceptual and interpretive aspects, based on five-month teaching internships via another university. The increase in teachers' noticing skills had a large effect size ($d = .79$) and was validated using a control group (Mertens & Gräsel, 2018; Mertens et al., 2018). Further evidence suggests that teaching internships benefit PSTs with weak teacher noticing skills, helping them to develop these skills (Orschulik, 2020; Stürmer et al., 2013).

In a similar study exploring the effects of two short-term (seven-week) practical activities at a German university, Weber et al. (2018) reported no significant development in the classroom management and general pedagogical knowledge-related facets of holistic professional vision for PSTs who participated in the practical activities without accompanying institutionalized reflection and feedback. In contrast, a group of PSTs who received peer and expert feedback on their videotaped lessons and gave feedback themselves significantly improved their skills, with a large effect size of $d = 1.10$ (Weber et al., 2018). Focused accompanying intervention to link theory and practice seems to support the development of teachers' noticing skills (Scholten & Orschulik, 2022; Stürmer et al., 2013).

However, until now the development of decision-making skills during teaching internships has not been investigated, and subject-specific insights in the noticing of mathematics teachers are lacking. Moreover, the influence of process variables (i.e., PSTs' individual teaching experiences and opportunities to learn within the internship) on the development of teacher noticing has not been examined before but are needed to assess benefits of teaching internships as an important part of initial teacher education and to provide first explanations for the development. Overall, previous results have been based solely on data collected at single universities, thus limiting their generalizability. Our study addressed this broad research gap.

6.2.2.4 Research questions

Based on current state-of-the-art research and the accompanying research desiderata, we aimed to investigate the reciprocal influences of the facets of teachers' noticing over the course of teaching internships and the influences of teaching internships and their process variables on the development of PSTs' teacher noticing. Therefore, we considered the following research questions:

1) To what extent do the facets of teachers' noticing at the beginning of teaching internships condition change over the course of the internships, considering autoregressive and cross-lagged effects?
2) To what extent do teaching internships affect changes in teachers' noticing skills?
3) To what extent do certain teaching internship process variables (i.e., learning time, teaching practice activities, and mentor support) influence teachers' noticing skills?

6.2.3 Methodological Approach

6.2.3.1 Context of the Study: Teacher Education in Germany

In this section, we briefly introduce teacher education in Germany and describe the characteristics of teaching internships to place the study in its context, to facilitate the understanding of our research design, methodology, and findings, and to enable researchers to connect our findings to the practical experiences in teacher education in their countries, as teaching internships have been identified as a crucial part of teacher education worldwide (Cabaroğlu & Öz, 2023; E. Cohen et al., 2013; Lawson et al., 2015).

In general, teacher education in Germany consists of three main phases, the first two constitute initial teacher education: (1) higher education at universities and pedagogical universities, (2) practical teacher training (known as preparatory service or induction) through specific state-based teacher seminars, and (3) elective professional development courses, which accompany professional practice as an in-service teacher following the first two phases of teacher education (Drahmann, 2020; Eckhardt, 2021).

In Germany, the exact design and organization of teacher education depends on the federal state (Bundesland) in which it is conducted, although shared norms and principles exist (Cortina & Thames, 2013). Several different types of schooling have been implemented for German secondary education (the focus of this paper): teaching at academic-track schools (Gymnasium), teaching at non-academic track schools, teaching at intermediate forms which combine academic and non-academic track education, teaching at vocational schools and special needs teaching (Cortina & Thames, 2013; Drahmann, 2020). For each type, PSTs must study two school subjects, with the exception of special needs education. University study is divided into four areas: subject matter, subject-related pedagogy, general pedagogy, and practical activities in school (Drahmann, 2020).

Due to the identified gap between theory and practice in teaching and teacher education (König & Rothland, 2018), connecting these practical activities with academic opportunities to learn has become increasingly important recently. Hence, school practical studies have been reorganized and reshaped in the wake of the Bologna reforms, that is, the transformation from a traditional state examination system to a bachelor-master-system comparable within Europe (Drahmann, 2020; Schubarth et al., 2012; Terhart, 2021). They are usually spread over several semesters of university teacher education depending on the federal state and aim to achieve several goals: professional orientation, competence improvement by linking theory and practice, and the development of teaching abilities (König, 2019).

In all federal states, PSTs undertake some extensive, long-term (i.e., lasting several months) practical activities in schools in the master's phase of their university teacher education to link theory with practice, which is often referred to as a teaching internship or (long-term) teaching practicum (Gröschner et al., 2015; Ulrich et al., 2020) and is, again, the focus of this paper. During their teaching internships, PSTs attend a school of their teaching type and participate in almost all aspects of daily school life. In particular, they are required to observe a certain number of lessons taught by in-service teachers and to teach lessons themselves under supervision. Mentors (i.e., in-service teachers at the school) support the PSTs in practical matters during their internships, while teacher educators from the university, and in some federal states, from the state-run teacher training seminars of the second phase of teacher education, facilitate the theory–practice linkage.

6.2.3.2 Sample

To address the aforementioned lack of studies based on multiple universities, the sample consisted of 175 secondary mathematics education PSTs from six universities (Hamburg, Cologne, Münster, Paderborn, Vechta, and Würzburg) in four German federal states who participated in an online survey before and after their (long-term) teaching internships between spring 2019 and fall 2021. At T1, 313 PSTs participated in the survey. The panel sample decreased with a panel attrition of 44% due to lack of participation at T2. Since the process variable data were collected at T2, these individuals were missing not only the T2 ability scores but also important predictor variables and were therefore excluded from the final panel sample. Data imputation was not possible because the missing predictor variables would have led to a circular process. Teaching internships were conducted at the master's level for all universities except Würzburg. Since initial teacher education in Bavaria is organized as a state examination with no division between bachelor's and master's programs, PSTs from Würzburg participated in the survey during their 5th or 6th semesters of study. The demographic sample statistics are presented in Table **Fehler! Verweisquelle konnte nicht gefunden werden** (Table 6.1).

Table 6.1 Descriptive characteristics of the total sample and the groups of different universities

University	*n*	Gender	Age	Semester[a]	Grade[b] (Abitur[c])	Teacher education program (%)			
		% female	*M (SD)*	*M (SD)*	*M (SD)*	I	II	III	IV
Hamburg	21	71.4	25.5 (2.2)	10.6 (3.8)	2.1 (0.60)	57.1	0	42.9	0
Cologne	49	61.2	25.9 (3.2)	9.7 (2.8)	1.8 (0.58)	14.3	83.7	0	2.0
Münster	64	56.3	25.8 (4.0)	9.7 (2.1)	2.1 (0.56)	26.6	67.2	0	6.3
Paderborn	5	80.0	23.2 (1.3)	8.6 (2.6)	2.1 (0.81)	20.0	80.0	0	0
Vechta	14	64.3	24.5 (2.2)	8.1 (2.3)	2.7 (0.32)	100.0	0	0	0
Würzburg	22	59.1	23.2 (1.2)	6.0 (1.0)	2.0 (0.65)	59.1	40.9	0	0
Total	175	61.1	25.3 (3.3)	9.1 (2.7)	2.0 (0.61)	36.6	55.4	5.1	2.9

Note. I ≙ non-academic track, II ≙ academic track (Gymnasium), III ≙ special needs education, IV ≙ vocational school

[a] Semester stands for the semester of study at the beginning of the teaching internship.

[b] German grades range from 1 (best) to 4 (pass).

[c] A German high school diploma that qualifies students for university admission.

6.2.3.3 Study Design, Assessment Instruments, and Scaling

Deploying a pretest–posttest design, we administered a survey to the participants via an online platform before their teaching internships (at T1) and after they completed their internships (at T2) in the frame of the TEDS-Validate-Transfer research project. Each questionnaire took approximately 90 minutes to complete. At T1, we obtained demographic data and measured teachers' noticing using a video-based instrument. At T2, the latter measure was repeated and supplemented with an assessment of process variables related to teaching internships: learning time, teaching practice activities, and mentor support. PSTs received a small financial reward for their participation. Data were collected and processed according to the requirements of the General Data Protection Regulation.

6.2.3.3.1 Instrument for Measuring Teacher Noticing

We employed an established instrument developed in the TEDS-FU study to measure teachers' noticing (Kaiser et al., 2015). It comprised three scripted (i.e., staged) video-vignettes, which ranged in length from 2.25 to 3.5 minutes and were presented in random order. The vignettes represented 9th to 10th grade lessons and covered a range of mathematical topics (e.g., surface and volume calculations, functions, and modeling) and teaching phases, such as introducing a mathematical task, working on it, and conducting a class discussion of the results (Kaiser et al., 2015). Prior to watching each video-vignette, the PSTs were provided with background information about the depicted class, pedagogical remarks, and the mathematical content to be addressed. They then viewed the corresponding video-vignette once and were asked to respond to 77 open-response and Likert-type rating-scale items that assessed their teacher noticing skills in perception ($n = 24$), interpretation ($n = 41$) or decision-making ($n = 11$) with either a mathematics pedagogical or general pedagogical focus.

The use of Likert scales, which is a widely used approach to measuring teacher noticing (Keppens et al., 2021; Weyers, König, Santagata, et al., 2023) and have already been used to evaluate teaching internships (Mertens & Gräsel, 2018), enabled a timesaving but accurate assessment of teachers' perceptions, interpretations, and decisions regarding distinct incidents. They were complemented by open-response items that allowed for the testing of more complex situations. The use of scripted video-vignettes and permitting only one-time access allowed for manageable and cognitively activating measurements that were strongly related to the classroom environment and realistic instructional situations (Piwowar et al., 2018; Santagata et al., 2021).

The rating-scale items comprised a statement and a four-point Likert response scale for assessing it. The example item in Fig. 6.2 was used to assess the perception of a specific classroom event. The interpretation item in Fig. 6.3, related to the mathematics pedagogical perspective, asked the participants to analyze a student's solutions to volume and surface calculations and to name indicators that support the hypothesis of the student's preference for formal mathematical approaches. The decision-making item in Fig. 6.4 regarding general pedagogy and inclusive education asked about dealing with class heterogeneity and possible changes in the course of instruction to better address this heterogeneity (Fig. 6.2–6.4).

The guiding question of the lesson is presented in written form.	fully correct	partially correct	partially incorrect	not correct at al
	o	o	o	o

Fig. 6.2 Rating scale item for the perception facet from a general pedagogical perspective

All items were scored dichotomously as correct (1) or incorrect (0), with the exception of five open-response partial-credit items coded 0 to 2 or 0 to 3. To evaluate the quality of the instrument and determine the correct and incorrect

What indicators in the video support the assumption that Kim prefers a formal mathematical approach rather than a descriptive one?

Name three.

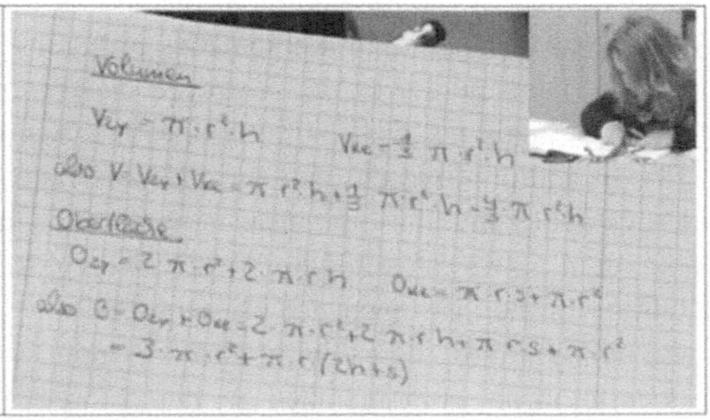

Fig. 6.3 Open-response item for the interpretation facet from a mathematics pedagogical perspective

Assume that the teacher feels that he is not addressing the ability heterogeneity of his class shown in the video with his instruction adequately.

Describe briefly,

a) at which point in the course of instruction

b) and in which way

you would modify the lesson shown - while retaining as many of its elements as possible - in order to better accommodate the different learning conditions within the class.

Fig. 6.4 Open-response item for the decision-making facet from a general pedagogical perspective

answers for the rating scale items, expert reviews were conducted during the development of the instrument (see Hoth et al., 2016, for details). Based on theoretical considerations and expert judgments, we created a detailed coding manual for scoring the open-response items. It consisted of comprehensive descriptions and multiple anchor examples to clarify the correct responses. The instrument's reliability was assessed using Cohen's kappa (J. Cohen, 1960) and double-coding for 20% of the responses to investigate intercoder agreement for each item. The resulting values ($\kappa_{mean} = .80$, $\kappa_{min} = .47$, $\kappa_{max} = 1.00$) indicated good overall intercoder reliability. We excluded five items with poor intercoder reliability due to a low frequency of correct responses. These items were discussed in detail by the raters and then coded by consensus.

The validity of the measurement instrument was examined through extensive workshops with experts in mathematics pedagogy and general pedagogy, focusing on the authenticity of the portrayed classroom situations and the adequacy of the test items, as well as curricular analyses of the content (Kaiser et al., 2015). In addition, we ensured the independence of the measurement from the video-vignettes and, thus, the measurement of the underlying construct (Blömeke, König, et al., 2015). A study by Weyers, König, Rott, et al. (2023) confirmed the suitability of the instrument for use with PSTs.

In our analyses, cases with 50% or more valid responses for each time of measurement were included in the dataset. We scaled the data collected using the teacher noticing measure with *ConQuest 5.28 software* (Adams et al., 1997–2023) with both measurement times combined using a three-dimensional Rasch model[4]. To estimate the item parameters, missing responses were considered not administered to include only valid answers. Missing responses were treated as incorrect for estimating a person's abilities. Weighted likelihood estimates (WLEs) were applied to create ability scores for perception, interpretation, and decision-making.

The item–total correlations ranged from .12 to .54, with an acceptable mean of .30. The weighted mean square (a component of the fit statistic that ideally yields a value of 1.0) varied in an acceptable range between .87 and 1.14, with an average of 1.00. The scales reached at least acceptable separation reliability, as shown in Table. 6.2. WLE reliability for decision-making was somewhat questionable, possibly due to the small total number of decision-making items associated with a complex construct and the variety of decision-making contexts represented in the instrument, which is a common problem when measuring decision-oriented constructs (Weyers et al., 2024). The attenuation-corrected latent correlations between the three facets were comparable to previous studies, with correlations of $r = .79$ between perception and interpretation, $r = .88$ between interpretation and decision-making, and $r = .50$ between perception and decision-making (Bastian et al., 2022).

To account for measurement invariance, we computed additional separate Rasch models for each measurement time. WLEs were then deployed to compare item difficulties at T1 and T2 (Bond et al., 2021). A correlation of $r = .97$ between T1 and T2 indicated a comparable measurement (see Fig. 6.5).

[4] The total number of missing responses was rather small (7.6%), so the person ability parameters could be estimated from the available data. Missing amounts per case ranged from 1–38% with a median of 3.9%.

Table 6.2 Scale reliability for all three facets of teachers' noticing

Reliability measure	Person separation reliability		
	Perception	Interpretation	Decision-making
EAP	.78	.87	.75
WLE	.69	.82	.48

Note: The expected A Posteriori (EAP) and weighted likelihood estimate (WLE) measures of person separation reliability were integral to the ConQuest software (Adams et al., 1997–2023) and can be interpreted like Cronbach's Alpha values.

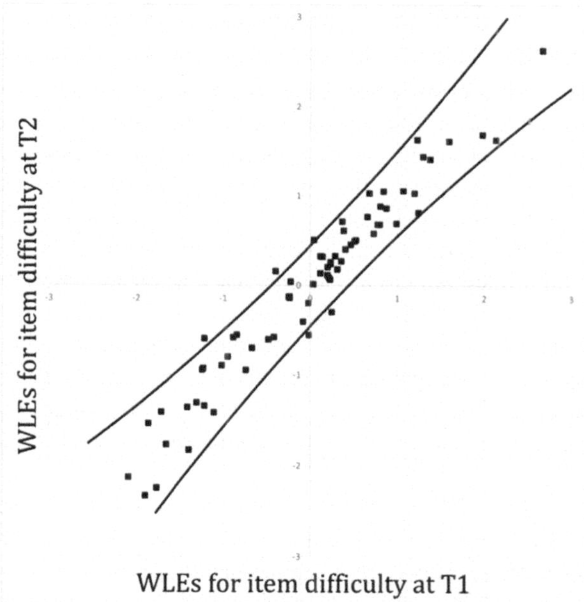

Fig. 6.5 Item difficulty invariance for T1 versus T2.

6.2.3.3.2 Instruments for Assessing the Teaching Internship process Variables

To gain insight into PSTs' individual experiences and the implementation of teaching internships, we adapted scales from the Learning to Practice study and the COACTIV-R study to assess (1) learning time (i.e., time spent on certain activities), (2) teaching practice activities, and (3) mentor support (König et al., 2014; Kunter et al., 2013). With these scales, we aimed to measure the number and quality of activities performed, as well as the perceived quality of mentor support, which previous studies have found to have an important influence on PST development, as described in the literature review. For the analyses, we transformed the learning time subscales into an interval scale measure. Teaching practices subscales were created using sum scores, mentor support subscales by mean scores. The subscales, sample items, and descriptive statistics are presented in Table 6.3. Internal consistency was at least acceptable (Table 6.3).

6.2.3.4 Data Analysis

To address the first research question concerning reciprocal effects of the three facets of teacher noticing, we computed autocorrelations and an autoregressive manifest model with a cross-lagged panel design using the WLEs with *Mplus 6.8* software (Muthén & Muthén, 1998–2021). Although the teacher education programs and, in particular, the teaching internships at all six universities in our sample shared key components, they were not, of course, completely similar. To account for differences and impacts of teacher educators, teaching type and modules and thus the stratified nature of the sample, we specified a stratification variable combining university and teaching types using the "type = complex" option.

We approached the second research question which addressed changes in PST's noticing skills using paired sample t-tests. We calculated the Cohen's *d* effect sizes for a within-subjects design (Lakens, 2013) to estimate the impact of teaching internships on teacher noticing and compare our results with previous findings.

Again, we used multiple regression analyses to investigate the third research question, considering the stratified sample. We conducted several analyses (for each set of subscales separately) to examine the effects of process variables on the ability scores for each teacher noticing facet at T2. Regression analyses were controlled for high school diploma grade, semester, and dichotomized teaching type (academic-track and vocational school vs. non-academic-track and special needs education), as well as the ability scores for the teacher noticing facets at T1.

Table 6.3 Overview of measures for the teaching internship process variables

Scale	Sample item	*n* items	*Median/ M (SD)*	α
Learning time[a]	As part of my internship, I …	6		.76
Lesson preparation	prepared school lessons	1	21–30	–
Observing instruction	observed school lessons	1	51–100	–
Supporting/Co-Teaching	assisted a teacher in the classroom (e.g., tutoring individual students or supporting group work)	1	31–50	–
Teaching with guidance	taught myself in the presence of a teacher	1	21–30	–
Teaching alone	taught myself without the presence of a teacher (e.g., took over substitute teaching)	1	1–10	–
Lesson follow-up	followed up on lessons.	1	1–10	–
Teaching practices[b]	I …	72		.90
Research-based learning	created an observation log	9	.35 (.22)	.59
Lesson planning	determined lesson content	12	.66 (.17)	.67
Teaching	discussed mistakes with students.	31	.59 (.19)	.86
Linking theories to situations	used educational theories to guide my actions in the internship.	11	.61 (.22)	.72
Reflecting on practice	followed up and analyzed my lessons on my own.	9	.55 (.21)	.58
Mentor support[c]		14		.91
Emotional	The teachers are understanding if things don't go so well during the internship.	4	5.03 (.96)	.84

(continued)

Table 6.3 (continued)

Scale	Sample item	*n* items	*Median/ M (SD)*	α
Informational	The teachers give me helpful tips on how to behave at school.	5	4.96 (.89)	.81
Instrumental	Teachers regularly share materials with me.	5	4.81 (.94)	.82

Note. For the learning time subscales, the median is given in hours. For teaching practice subscales, the percentage mean is reported because the number of items per subscale varied. For the mentor support subscales, the mean was taken from a six-point rating scale.
[a] For each item, the time commitment was rated on a seven-point rating scale ranging from "No" (1) to "Yes, more than 100 hours" (7).
[b] For each item, participants indicated whether or not they had performed the described activity.
[c] For each item, participants rated the given statement on a six-point rating scale ranging from "does not apply at all" (1) to "fully applies" (6).

6.2.4 Results

We now present the results in three subsections, each of which addresses one of our research questions, before discussing the results in the next section.

6.2.4.1 Reciprocal Effects of the Facets of Teachers' Noticing during Teaching Internships (First Research Question)

To answer our first research question, we conducted cross-lagged panel analyses. The stability of the teacher noticing facets was investigated using autocorrelations. Positive medium to strong autocorrelations for perception (r = .50, p < .001), interpretation (r = .62, p < .001) and decision-making (r = .52, p < .001) indicated stable parts of these constructs, but also emphasized intrapersonal variation and, thus, change during the teaching internships[5]. We applied a cross-lagged panel model to assess the predictive quality of the three facets at T1 for the facets at T2 (see Fig. 6.6). The means and standard deviations for all ability scores are presented in Table 6.4, and the correlations between all facets of teachers' noticing are presented in Table 6.5. The correlations suggested possible cross-lagged effects between the three facets. Measures of reliability have already been discussed (see Table 6.2).

[5] Values above .10/.30/.50 indicate small/medium/strong effects, respectively (J. Cohen, 1988).

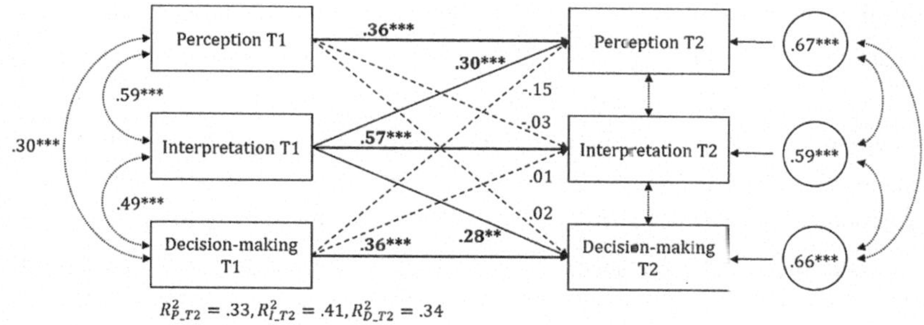

Note * ≙ $p < .05$, ** ≙ $p < .01$, *** ≙ $p < .001$. $R^2_{P\_T2}/R^2_{I\_T2}/R^2_{D\_T2}$ ≙ coefficient of determination for perception at T2/interpretation at T2/decision-making at T2, respectively. In the analysis, we controlled for high school diploma grade, semester, and dichotomized teaching type. The path coefficients display the standardized results. Significant regression coefficients are shown in bold for ease of reading. Values in ovals on the right indicate unexplained variance.

Fig. 6.6 Cross-lagged panel model for the facets of teachers' noticing. $R^2_{P\_T2}/R^2_{I\_T2}/R^2_{D\_T2}$

The auto-lagged path coefficients showed significant moderate predictive power for perception ($\beta = .36$, $p < .001$) and decision-making ($\beta = .36$, $p < .001$), and a strong effect for interpretation ($\beta = .57$, $p < .001$). The cross-lagged path coefficients demonstrated that only interpretation had an effect on the other facets, since both paths from interpretation at T1 to perception at T2 ($\beta = .30$, $p = .001$) and decision-making at T2 ($\beta = .28$, $p = .002$) were significant, but not the reverse cross-lagged paths. This indicated a prominent role for interpretation in the development of all three facets. No cross-lagged effect was observed between perception and decision-making. For all three facets, the model explained a significant amount of the variance (Table 6.4 and 6.5).

Table 6.4 Mean scores at each measurement point for the teachers' noticing facets

Variable	T1		T2		t (174)	*p*	Cohen's *d*
	M	*SD*	*M*	*SD*			
Perception	48.47	10.31	51.53	9.47	4.53	< .001	.31
Interpretation	48.50	9.92	51.50	9.88	4.30	< .001	.39
Decision-making	48.50	10.12	51.50	9.68	4.46	< .001	.31

Note: To facilitate reading and interpretation, we linearly transformed the overall ability estimates for all facets obtained from the scaling to a mean of 50 and a standard deviation of 10.

Table 6.5 Correlations between teachers' noticing facets at T1 and T2

Variable	Perception T2	Interpretation T2	Decision-making T2
Perception T1	.504***	.341***	.304***
Interpretation T1	.457***	.617***	.489***
Decision-making T1	.124	.330***	.518***

Note. * ≙ $p < .05$, ** ≙ $p < .01$, *** ≙ $p < .001$.

6.2.4.2 Effects of Teaching Internships on Teachers' Noticing (Second Research Question)

To answer the second research question, we compared the measurements of individual teachers' noticing skills before and after the teaching internships using t-tests to explore the effects of the practical activities (see Table 6.4). Significant increases were found for all three facets of teachers' noticing over the course of the teaching internships. The effect sizes, expressed as Cohen's *d*, revealed small effects for all three facets, with a higher effect size for interpretation.

6.2.4.3 Effects of process Variables on Teachers' Noticing (Third Research Question)

The third research question was answered using multiple regression analyses to investigate the effects of the teaching internship process variables on the development of PSTs' teacher noticing. For each scale (i.e., learning time, teaching practice activities, and mentor support) and for each noticing facet a model was calculated using all subscales of the scale as predictors. The results of these analyses are illustrated in Table 6.6, Table 6.7, and Table 6.8. They showed the significant influences of process variables for only a few subscales. A significant proportion of the variance was explained.

Table 6.6 Multiple regression models for the effects of learning time

Variable	Perception T2			Interpretation T2			Decision-making T2		
	β	t	p	β	t	p	β	t	p
Lesson preparation	−.06	−.50	.624	−.06	−.50	.624	−.21*	−2.06	.039
Classroom observation	.01	.10	.921	−.08	−1.01	.312	−.02	−.28	.776
Lessons together with a teacher	.00	.04	.969	−.01	−.08	.933	−.03	−.39	.694
Own lessons with the presence of a teacher	−.05	−.48	.635	.01	.11	.912	.03	.28	.783
Own lessons without the presence of a teacher	−.04	−.88	.378	−.03	−.91	.362	−.04	−.82	.414
Lesson follow-up	−.05	−.89	.392	.03	.51	.614	.19**	2.73	.006
R^2	.35			.42			.38		

Note. * ≙ $p < .05$, ** ≙ $p < .01$, *** ≙ $p < .001$, β ≙ standardized regression coefficient. In the analyses, we controlled for high school diploma grade, semester, and dichotomized teaching type. We computed separate models for each criterion.

For learning time, there was only a positive effect of lesson follow-up on decision-making (β = .19, p = .006), while lesson preparation negatively influenced decision-making (β = −.21, p = .039) Table 6.6. For the teaching practice activities, linking theories to specific situations significantly predicted perception and interpretation, with positive regression coefficients of β = .19, p = .014 and β = .25, p < .001, respectively. In contrast, teaching (i.e., performing situations) showed a negative effect on perception (β = −.26, p = .001) but no effects on interpretation and decision-making (Table 6.7).

Emotional support provided by PSTs' mentors had a significant positive effect on decision-making (β = .27, p = .002). In contrast, instrumental support, such as providing teaching materials, negatively affected interpretation (β = −.14, p = .048) and decision-making skills (β = −.15, p = .046) (Table 6.8).

Table 6.7 Multiple regression models for the effects of teaching practice activities conducted

Variable	Perception T2			Interpretation T2			Decision-making T2		
	β	t	p	β	t	p	β	t	p
Research-based learning	.03	.35	.724	−.06	−.81	.418	−.04	−.59	.553
Lesson planning	.04	.64	.525	−.06	−.82	.410	−.01	−.12	.907
Teaching	−.26**	−3.38	.001	−.02	−.19	.846	−.05	−.65	.516
Linking theories to situations	.19*	2.46	.014	.25***	3.58	<.001	.14	1.75	.081
Reflecting on practice	−.02	−.26	.792	−.10	−1.19	.236	−.03	−.29	.781
R^2	.39			.46			.36		

Note. * ≙ $p < .05$, ** ≙ $p < .01$, *** ≙ $p < .001$, β ≙ standardized regression coefficient. In the analyses, we controlled for high school diploma grade, semester, and dichotomized teaching type. We computed separate models for each criterion.

Table 6.8 Multiple regression models for the effects of mentor support

Variable	Perception T2			Interpretation T2			Decision-making T2		
	β	t	p	β	t	p	β	t	p
Emotional	−.04	−.32	.750	.16	1.90	.058	.27**	3.10	.002
Informational	−.01	−.08	.940	−.13	−1.43	.152	−.11	−1.54	.123
Instrumental	.03	.31	.754	−.14*	−1.98	.048	−.15*	−2.00	.046
R^2	.33			.44			.38		

Note. * ≙ $p < .05$, ** ≙ $p < .01$, *** ≙ $p < .001$, β ≙ standardized regression coefficient. In the analyses, we controlled for high school diploma grade, semester, and dichotomized teaching type. We computed separate models for each criterion.

6.2.5 Discussion and Limitations of the Study

In this study, we aimed to examine the reciprocal effects of the facets of teachers' noticing during teaching internships, together with the effects of the internships and their implementation on the development of PSTs' teacher noticing. We now discuss the results for each research question and the limitations of this study.

6.2.5.1 Discussion and Implications

6.2.5.1.1 Reciprocal Effects of the Facets of Teachers' Noticing during Teaching Internships

We examined teachers' noticing and its three facets (perception, interpretation, and decision-making) in the course of teaching internships. The results revealed that the three facets improved significantly during the internships, and abilities at T1 significantly predicted ability scores at T2, particularly for interpretation. This is consistent with a study by Jong et al. (2021), who reported similar effects for the noticing facets of perception and interpretation. However, their regression coefficients were smaller than those in our study, and no significant prediction was found for decision-making. Since Jong et al. (2021) investigated the development of teachers' noticing during a university course that did not include practical activities in schools, this suggests that practical field experiences play an important role for facilitating PSTs' development of teacher noticing skills.

Furthermore, the cross-lagged analysis highlighted interpretation skills as vital for the development of teachers' noticing, since interpretation had cross-lagged effects on perception and decision-making, but no reverse paths were significant. This confirms previous findings regarding the influence of interpretation on perception (Jong et al., 2021; Superfine et al., 2017) and provides further insight into the development of decision-making skills. The results of the cross-lagged analysis may also indicate a causal effect of interpretation on the development of perceptual and decision-making skills.

Additionally, the critical role of interpretation in our data underscores the conceptualization of teachers' noticing as a knowledge-based construct (Sherin, 2007), challenging prior conceptualizations implying a more linear learning path of perception followed by interpretation and then decision-making (van Es, 2011). Our findings suggest that the ability to interpret classroom events and student thinking seem to be key to developing all three facets of teachers' noticing and, thus, to perceiving relevant details and making productive decisions in the classroom. Initial learning to interpret might facilitate perception and decision-making in the classroom (Superfine et al., 2017). Our findings also suggest that interpretation skills can facilitate the application of professional knowledge in situation-specific school contexts. Hence, it is crucial to provide PSTs with sufficient opportunities to learn interpreting before, during, and after their teaching internships.

6.2.5.1.2 Effects of Teaching Internships on Teachers' Noticing

We investigated the impact of the teaching internships on perception, interpretation, and decision-making by comparing the PSTs' pre- and posttest ability scores. The results showed significant increases for all three facets. The changes observed corroborated similar findings by Stürmer et al. (2013) and Mertens and Gräsel (2018), again supporting the important role of teaching internships in fostering the development of PSTs' teachers' noticing during their university education and complementing these studies with a subject-specific perspective on teacher noticing. Moreover, previous studies have not reported this improvement in decision-making skills; thus, our results make a recent contribution to the literature. However, the effect sizes in our study were somewhat smaller than expected based on previous studies (Mertens & Gräsel, 2018; Weber et al., 2018) and indicated only small but significant effects. This might be explained by the weaker connection between university education and teaching internships in some parts of North Rhine-Westphalia, where a significant part of the participants studied, offering fewer opportunities to relate theory to practice (Doll et al., 2018). Previous studies have suggested that the degree of theory–practice linkage may have an impact on the effect size of changes in teachers' noticing over the course of teaching internships (Weber et al., 2018).

6.2.5.1.3 Effects of Process Variables on Teachers' Noticing

We explored the influences of the teaching internship process variables and, thus, the organization and implementation of internships for the individual PSTs using regression analyses, which showed effects for only some of the variables, particularly positive influences of making connection between theory and practice and emotional mentor support. This raises questions about the necessity for and current form of some features of teaching internships, such as lesson planning, which, surprisingly, showed no effects or negative effects on the three facets of teachers' noticing.

The amount of time spent on lesson preparation had a negative effect on decision-making, which may be explained by the PSTs' lack of lesson planning skills; that is, the PSTs were unable to make sufficient use of this time due to their lack of knowledge about lesson planning (König, Cammann, et al., 2022). On the other hand, this result may indicate that PSTs who prepared longer or more extensively for lessons were more restricted in their expectations about the decisions to be made in class, had fewer opportunities to practice a variety of decisions, and thus fewer opportunities to develop their decision-making skills. Furthermore, teaching (e.g., teaching one's own lessons and assisting a teacher in co-teaching) had no positive effect on teachers' noticing. This is particularly interesting since PSTs often request more of these activities in their university education (Hascher, 2014; Ulrich et al., 2020) and engage extensively in these activities during their teaching internships (see Table 6.3; König, Darge, Kramer, et al., 2018). Time spent on teaching even had a negative impact on perception. Hence, participation in classroom teaching alone does not seem to be sufficient for developing teachers' noticing competence, complementing prior results that demonstrated the risks of adopting established school practices from in-service teacher without reflecting on them (Chitpin et al., 2008; Özgün-Koca & İlhan Şen, 2006). PSTs who focus their teaching internships predominately on performing situations may have less time for processing and reflection and may therefore adopt established school practices without questioning them. Extended practical learning opportunities thus do not automatically lead to increases in teacher competence, as empirical findings from TEDS-M have indicated for teacher education in Germany and the United States (König & Blömeke, 2012).

However, making connections between theory and experience in the internships and reflecting on practice appear to facilitate the development of teachers' noticing, as linking theories to situations significantly predicted perception and interpretation, and time spent on lesson follow-up predicted decision-making. These findings agree with those of Weber et al. (2018, 2020) and Stürmer et al. (2013), who described growth in teachers' noticing skills, particularly in the context of accompanying reflection-oriented and analysis-oriented activities, and also with similar results for the development of professional knowledge (Schlag & Glock, 2019). Furthermore, the results also accord with the analyses of König and Blömeke (2012), who reported higher levels of general pedagogical knowledge for PSTs who focused on reflection rather than on teaching alone. Establishing connections between theory and the field experiences of PSTs (e.g., through reflection) seems to be a decisive factor in the development of teachers' noticing during teaching internships.

Mentor support has proved to be highly important for PSTs during their teaching internships (García-Lázaro et al., 2022; Hudson & Millwater, 2008). In this study, only emotional support facilitated PSTs' development, and only in terms of decision-making. The opportunity to talk with experienced teachers, express their concerns and uncertainties, and receive encouragement seems to help PSTs make in-the-moment decisions in classroom situations. In contrast, instrumental support (i.e., the provision of teaching methods and materials, such as worksheets) had a negative relationship with interpretation and decision-making. Again, a possible explanation may be that PSTs who receive more instrumental support adopt teaching styles and methods in an unreflective manner and have fewer opportunities to connect educational theories independently with their applications in practice. On the other hand, PSTs with higher interpretive and decision-making skills may require less instrumental support.

6.2.5.2 Limitations

The limitations of this study should be considered. First, the results presented herein were derived from convenience sampling. Consequently, generalizations should only be made with caution. Group effects of the six-university sample were controlled for in the analyses using stratification variables. However, this approach may not have accounted for all effects of the different locations and may potentially have overlooked additional group-specific effects. In addition, for organizational reasons, it was not possible to establish control groups to evaluate the influence of the teaching internships against a group that did not undertake internships and to control for memory effects. However, this only influenced the overall effect of the internships on the three facets of teachers' noticing for the second research question.

Moreover, the design of the study included only two measurement time points, which provided a rather rough representation of the development of teachers' noticing over the course of the teaching internships. Research with more measurements at shorter time intervals is needed to gain more insight into the development of teachers' noticing and, in particular, into the optimal length of an internship, since saturation effects may occur during this activity.

Process variables (i.e., learning time, teaching practice activities, and mentor support) were measured only through self-report instruments, which may have created bias in the dataset. Third-party assessments by mentors or researchers, for example, should be used to complement our findings. Furthermore, our test instrument assessed only a subset of teachers' professional competencies; thus, the development of other competence facets may have been overlooked, and more comprehensive survey designs are needed to consider knowledge, beliefs, and situation-specific and/or performance-related facets together and to obtain a deeper understanding of the effects of teaching internships on PSTs.

Finally, some parts of the data collection took place during the COVID-19 pandemic. This influenced the organization of the teaching internships in ways that differed among schools. Our items for the internship process variables specifically included participation in online and distance learning and teaching, which provided some control for the changed conditions. Nevertheless, the pandemic may have had some unknown effects on our study.

6.2.6 Conclusions

Teaching internships have become an increasingly important, but also challenging and expensive, part of initial teacher education for the development of PSTs' professional competence, which calls for empirical studies in this area and particularly quantitative analyses towards potential effects on PST learning outcomes (Lawson et al., 2015; König & Rothland, 2018; Terhart, 2021; Ulrich et al., 2020). This study addressed research gaps towards the development of teachers' noticing over the course of teaching internships, in particular providing insight for the first time into the development of subject-specific teacher noticing skills and particularly decision-making skills, the effects of process variables of teaching internships, and the reciprocal effects of facets of teacher noticing. As the results are based on an established teacher noticing instrument and a framework that combines general pedagogical as well as mathematics pedagogical perspectives, and the core characteristics of German teaching internships are comparable to other international formats, they promise to be meaningful in contexts other than in this study. The results can inform the structure and organization of teaching internships and field experiences in initial teacher education internationally, as well as providing insights into the development of teacher noticing on a theoretical level in general.

The development of interpreting as a key skill for enhancing teachers' noticing and applying knowledge learned at the university to the classroom was shown to be of great importance in this study. Therefore, we propose to focus more strongly on this facet to prepare PSTs for and accompany changes in teachers' noticing during teaching internships. The initial fostering of interpretation may reduce later cognitive load and make it easier for PSTs to learn perception and decision-making in a meaningful way. Since few variables explained interpretation skills in our analyses, further research is needed to explore how interpretation skills can be fostered and what variables might explain the development of these skills. Weyers, König, Rott, et al.'s (2023) results suggest effects on interpretation skills of average high-school diploma grade and, thus, academic capability, as well as opportunities to learn from university education.

Regarding the structure of teaching internships, our findings suggest a need to strengthen theory–practice linkage activities in the practice of teaching internships. Overall, our study suggests that teaching internships and teaching practice activities have the potential to foster teachers' noticing as a central facet of future professional practice and promote the connection between practice, teacher noticing, and academic knowledge.

References

Adams, R. J., Wu, M., Macaskill, G., Haldane, S., Sun, X. X., Cloney, D., & Berezner, A. (1997–2023). *ConQuest* (Version 5.28) [Computer software]. Australian Council for Educational Research.

Amador, J. M., Bragelman, J., & Superfine, A. C. (2021). Prospective teachers' noticing: A literature review of methodological approaches to support and analyze noticing. *Teaching and Teacher Education, 99*, 1–16. https://doi.org/10.1016/j.tate.2020.103256

Bastian, A., Kaiser, G., Meyer, D., Schwarz, B., & König, J. (2022). Teachers' noticing and its growth toward expertise: An expert–novice comparison with pre-service and in-service secondary mathematics teachers. *Educational Studies in Mathematics, 110*(2), 205–232. https://doi.org/10.1007/s10649-021-10128-y

Berliner, D. C. (1988). *The development of expertise in pedagogy.* American Association of Colleges for Teacher Education, Washington, D.C.

Blömeke, S., Gustafsson, J. E., & Shavelson, R. J. (2015). Beyond dichotomies: Competence viewed as a continuum. *Zeitschrift Für Psychologie / Journal of Psychology, 223*(1), 3–13. https://doi.org/10.1027/2151-2604/a000194

Blömeke, S., Jentsch, A., Ross, N., Kaiser, G., & König, J. (2022). Opening up the black box: Teacher competence, instructional quality, and students' learning progress. *Learning and Instruction, 79*, 1–11. https://doi.org/10.1016/j.learninstruc.2022.101600

Blömeke, S., König, J., Suhl, U., Hoth, J., & Döhrmann, M. (2015). Wie situationsbezogen ist die Kompetenz von Lehrkräften? Zur Generalisierbarkeit der Ergebnisse von videobasierten Performanztests. *Zeitschrift Für Pädagogik, 61*(3), 310–327.

Bond, T. G., Yan, Z., & Heene, M. (2021). *Applying the Rasch model: Fundamental measurement in the human sciences.* Routledge Taylor & Francis Group.

Cabaroğlu, N., & Öz, G. (2023). Practicum in ELT: a systematic review of 2010–2020 research on ELT practicum. *European Journal of Teacher Education*, 1–20. https://doi.org/10.1080/02619768.2023.2242577

Caires, S., & Almeida, L. S. (2005). Teaching practice in Initial Teacher Education: its impact on student teachers' professional skills and development. *Journal of Education for Teaching, 31*(2), 111–120. https://doi.org/10.1080/02607470500127236

Chitpin, S., Simon, M., & Galipeau, J. (2008). Pre-service teachers' use of the objective knowledge framework for reflection during practicum. *Teaching and Teacher Education, 24*(8), 2049–2058. https://doi.org/10.1016/j.tate.2008.04.001

Cohen, J. (1960). A coefficient of agreement for nominal scales. *Educational and Psychological Measurement, 20*(1), 37–46. https://doi.org/10.1177/001316446002000104

Cohen, J. (1988). *Statistical power analysis for the behavioral sciences.* Lawrence Erlbaum Associates.

Cohen, E., Hoz, R., & Kaplan, H. (2013). The practicum in preservice teacher education: a review of empirical studies. *Teaching Education*, 24(4), 345–380. https://doi.org/10.1080/10476210.2012.711815

Cortina, K. S., & Thames, M. H. (2013). "Teacher Education in Germany," in *Cognitive Activation in the Mathematics Classroom and Professional Competence of Teachers*, eds. M. Kunter, J. Baumert, W. Blum, U. Klusmann, S. Krauss, & M. Neubrand (Springer New York), 49–62. https://doi.org/10.1007/978-1-4614-5149-5_3

Dindyal, J., Schack, E. O., Choy, B. H., & Sherin, M. G. (2021). Exploring the terrains of mathematics teachers' noticing. *ZDM—Mathematics Education, 53*(1), 1–16. https://doi.org/10.1007/s11858-021-01249-y

Doll, J., Jentsch, A., Meyer, D., Kaiser, G., Kaspar, K., & König, J. (2018). Zur Nutzung schulpraktischer Lerngelegenheiten an zwei deutschen Hochschulen: lernprozessbezogene Tätigkeiten angehender Lehrpersonen in Masterpraktika. *Lehrerbildung Auf Dem Prüfstand, 11*(1), 24–45.

Dominguez, H. (2019). Theorizing reciprocal noticing with non-dominant students in mathematics. *Educational Studies in Mathematics,* 102(1), 75–89. https://doi.org/10.1007/s10649-019-09896-5

Drahmann, M. (2020). "Teacher Education in Germany: A Holistic View of Structure, Curriculum, Development and Challenges," in *Teacher Education in the Global Era: Perspectives and Practices*, ed. K. Pushpanadham (Springer Singapore), 13–31. https://doi.org/10.1007/978-981-15-4008-0_2

Dreher, A., & Kuntze, S. (2015). Teachers' professional knowledge and noticing: The case of multiple representations in the mathematics classroom. *Educational Studies in Mathematics, 88*(1), 89–114. https://doi.org/10.1007/s10649-014-9577-8

Eckhardt, T. (2021). *The education system in the Federal Republic of Germany 2018/2019: A description of the responsibilities, structures and developments in education policy for the exchange of information in Europe.* https://www.kmk.org/fileadmin/Dateien/pdf/Eurydice/Bildungswesen-engl-pdfs/dossier_en_ebook.pdf

Fernandez, C., & Choy, B. H. (2019). Theoretical lenses to develop mathematidcs teacher noticing. In S. Llinares & O. Chapman (Eds.), *International Handbook of Mathematics Teacher Education: Volume 2: Tools and Processes in Mathematics Teacher Education* (pp. 337–360). Brill. https://doi.org/10.1163/9789004418967

Festner, D., Gröschner, A., Goller, M., & Hascher, T. (2020). "Lernen zu Unterrichten—Veränderungen in den Einstellungsmustern von Lehramtsstudierenden während des Praxissemesters im Zusammenhang mit mentorieller Lernbegleitung und Kompetenzeinschätzung," in *Praxissemester im Lehramtsstudium in Deutschland: Wirkungen auf Studierende*, eds. I. Ulrich & A. Gröschner (Springer Fachmedien), 209–241.

García-Lázaro, I., Colás-Bravo, M. P., & Conde-Jiménez, J. (2022). The Impact of Perceived Self-Efficacy and Satisfaction on Preservice Teachers' Well-Being during the Practicum Experience. *Sustainability, 14*(16), 10185. https://doi.org/10.3390/su141610185

Gold, B., & Holodynski, M. (2017). Using digital video to measure the professional vision of elementary classroom management: Test validation and methodological challenges. *Computers & Education, 107*, 13–30. https://doi.org/10.1016/j.compedu.2016.12.012

Goodwin, C. (1994). Professional vision. *American Anthropologist, 96*(3), 606–633.

Gröschner, A., Müller, K., Bauer, J., Seidel, T., Prenzel, M., Kauper, T., & Möller, J. (2015). Praxisphasen in der Lehrerausbildung—Eine Strukturanalyse am Beispiel des gymnasialen Lehramtsstudiums in Deutschland. *Zeitschrift für Erziehungswissenschaft, 18*(4), 639–665. https://doi.org/10.1007/s11618-015-0636-4

Gröschner, A., Schmitt, C., & Seidel, T. (2013). Veränderung subjektiver Kompetenzeinschätzungen von Lehramtsstudierenden im Praxissemester. *Zeitschrift Für Pädagogische Psychologie, 27*(1-2), 77–86. https://doi.org/10.1024/1010-0652/a000090

Hascher, T. (2014). Forschung zur Wirksamkeit der Lehrerbildung. In E. Terhart, H. Bennewitz, & M. Rothland (Eds.), *Handbuch der Forschung zum Lehrerberuf* (pp. 418–440). Waxmann.

Hoth, J., Schwarz, B., Kaiser, G., Busse, A., König, J., & Blömeke, S. (2016). Uncovering predictors of disagreement: ensuring the quality of expert ratings. *ZDM Mathematics Education*, 48(1-2), 83–95. https://doi.org/10.1007/s11858-016-0758-z

Hudson, P., & Millwater, J. (2008). Mentors' Views about Developing Effective English Teaching Practices. *Australian Journal of Teacher Education, 33*(5). https://doi.org/10.14221/ajte.2008v33n5.1

Jacobs, V. R., Lamb, L. L. C., & Philipp, R. A. (2010). Professional noticing of children's mathematical thinking. *Journal for Research in Mathematics Education, 41*(2), 169–202. www.jstor.org/stable/20720130

Jong, C., Schack, E. O., Fisher, M. H., Thomas, J., & Dueber, D. (2021). What role does professional noticing play? Examining connections with affect and mathematical knowledge for teaching among preservice teachers. *ZDM—Mathematics Education, 53*(1), 151–164. https://doi.org/10.1007/s11858-020-01210-5

Kaiser, G., Blömeke, S., König, J., Busse, A., Döhrmann, M., & Hoth, J. (2017). Professional competencies of (prospective) mathematics teachers—cognitive versus situated approaches. *Educational Studies in Mathematics, 94*(2), 1–22. https://doi.org/10.1007/s10649-016-9713-8

Kaiser, G., Busse, A., Hoth, J., König, J., & Blömeke, S. (2015). About the complexities of video-based assessments: Theoretical and methodological approaches to overcoming shortcomings of research on teachers' competence. *International Journal of Science and Mathematics Education, 13*(2), 369–387. https://doi.org/10.1007/s10763-015-9616-7

Keppens, K., Consuegra, E., Maeyer, S. de, & Vanderlinde, R. (2021). Teacher beliefs, self-efficacy and professional vision: disentangling their relationship in the context of inclusive teaching. *Journal of Curriculum Studies, 53*(3), 314–332. https://doi.org/10.1080/00220272.2021.1881167

König, J. (2019). "Empirische Befunde zu Effekten von Praxisphasen in der Lehrerausbildung," in *IFS-Bildungsdialoge: Band 3. Lehrerbildung—Potentiale und Herausforderungen in den drei Phasen*, eds. N. McElvany, F. Schwabe, W. Bos, & H. G. Holtappels (Waxmann), 29–52.

König, J., & Blömeke, S. (2012). Future teachers' general pedagogical knowledge from a comparative perspective: does school experience matter? *ZDM Mathematics Education, 44*(3), 341–354. https://doi.org/10.1007/s11858-012-0394-1

König, J., Cammann, F., Bremerich-Vos, A., & Buchholtz, C. (2022). Unterrichtsplanungskompetenz von (angehenden) Deutschlehrkräften der Sekundarstufe: Testkonstruktion und Validierung. *Zeitschrift für Erziehungswissenschaft, 25*(4), 869–894. https://doi.org/10.1007/s11618-022-01113-z

König, J., Darge, K., Klemenz, S., & Seifert, A. (2018). "Pädagogisches Wissen von Lehramtsstudierenden im Praxissemester: Ziel schulpraktischen Lernens?," in *Learning to practice, learning to reflect? Ergebnisse aus der Längsschnittstudie LtP zur Nutzung und Wirkung des Praxissemesters in der Lehrerbildung*, eds. J. König, M. Rothland, & N. Schaper (Springer VS), 287–323.

König, J., Darge, K., Kramer, C., Ligtvoet, R., Lünnemann, M., Podlecki, A.-M., & Strauß, S. (2018). "Das Praxissemester als Lerngelegenheit: Modellierung lernprozessbezogener

Tätigkeiten und ihrer Bedingungsfaktoren im Spannungsfeld zwischen Universität und Schulpraxis," in *Learning to practice, learning to reflect? Ergebnisse aus der Längsschnittstudie LtP zur Nutzung und Wirkung des Praxissemesters in der Lehrerbildung*, eds. J. König, M. Rothland, & N. Schaper (Springer VS), pp. 87–114.

König, J., & Rothland, M. (2018). "Das Praxissemester in der Lehrerbildung: Stand der Forschung und zentrale Ergebnisse des Projekts Learning to Practice," in *Learning to practice, learning to reflect? Ergebnisse aus der Längsschnittstudie LtP zur Nutzung und Wirkung des Praxissemesters in der Lehrerbildung*, eds. J. König, M. Rothland, & N. Schaper (Springer VS), 1–62. https://doi.org/10.1007/978-3-658-19536-6_1

König, J., Santagata, R., Scheiner, T., Adleff, A.-K., Yang, X., & Kaiser, G. (2022). Teachers' noticing: A systematic literature review of conceptualizations, research designs, and findings on learning to notice. *Educational Research Review, 36.* https://doi.org/10.1016/j.edurev.2022.100453

König, J., Tachtsoglou, S., Darge, K., & Lünnemann, M. (2014). Zur Nutzung von Praxis: Modellierung und Validierung lernprozessbezogener Tätigkeiten von angehenden Lehrkräften im Rahmen ihrer schulpraktischen Ausbildung. Zeitschrift für Bildungsforschung, 4(1), 3–22. https://doi.org/10.1007/s35834-013-0084-2

Kunter, M., Baumert, J., Blum, W., Klusmann, U., Krauss, S., & Neubrand, M. (Eds.). (2013). *Cognitive Activation in the Mathematics Classroom and Professional Competence of Teachers*. Springer New York.

Lakens, D. (2013). Calculating and reporting effect sizes to facilitate cumulative science: A practical primer for t-tests and ANOVAs. *Frontiers in Psychology, 4*, 863. https://doi.org/10.3389/fpsyg.2013.00863

Lawson, T., Çakmak, M., Gündüz, M., & Busher, H. (2015). Research on teaching practicum—a systematic review. *European Journal of Teacher Education, 38*(3), 392–407. https://doi.org/10.1080/02619768.2014.994060

Lenhard, W., & Lenhard, A. (2016 -2022). *Berechnung von Effektstärken*. Psychometrica. https://doi.org/10.13140/RG.2.2.17823.92329

Louie, N., Adiredja, A. P., & Jessup, N. (2021). Teachers' noticing from a sociopolitical perspective: The FAIR framework for anti-deficit noticing. *ZDM—Mathematics Education, 53*(1), 95–107. https://doi.org/10.1007/s11858-021-01229-2

Mason, J. (2002). *Researching your own practice: The discipline of noticing*. Routledge, London, New York. http://oro.open.ac.uk/790/

Mertens, S., & Gräsel, C. (2018). Entwicklungsbereiche bildungswissenschaftlicher Kompetenzen von Lehramtsstudierenden im Praxissemester. *Zeitschrift für Erziehungswissenschaft, 21*(6), 1109–1133. https://doi.org/10.1007/s11618-018-0825-z

Mertens, S., Schlag, S., & Gräsel, C. (2018). Die Bedeutung der Berufswahlmotivation, Selbstregulation und Kompetenzselbsteinschätzungen für das bildungswissenschaftliche Professionswissen und die Unterrichtswahrnehmung angehender Lehrkräfte zu Beginn und am Ende des Praxissemesters. *Lehrerbildung Auf Dem Prüfstand, 11*(1), 66–84.

Metsäpelto, R.-L., Poikkeus, A.-M., Heikkilä, M., Husu, J., Laine, A., Lappalainen, K., Lähteenmäki, M., Mikkilä-Erdmann, M., Warinowski, A., Iiskala, T., Hangelin, S., Harmoinen, S., Holmström, A., Kyrö-Ämmälä, O., Lehesvuori, S., Mankki, V., & Suvilehto, P. (2021). A multidimensional adapted process model of teaching. *Educational Assessment, Evaluation and Accountability*, 1–30. https://doi.org/10.1007/s11092-021-09373-9

Muthén, L. K., & Muthén, B. O. (1998–2021). *Mplus* (Version 8.6) [Computer software]. Muthén & Muthén.

Ng, W., Nicholas, H., & Williams, A. (2010). School experience influences on pre-service teachers' evolving beliefs about effective teaching. *Teaching and Teacher Education, 26*(2), 278–289. https://doi.org/10.1016/j.tate.2009.03.010

Orschulik, A. B. (2020). *Entwicklung der professionellen Unterrichtswahrnehmung: Eine Studie zur Entwicklung Studierender in universitären Praxisphasen.* [dissertation thesis]. Springer Spektrum. https://doi.org/10.1007/978-3-658-33931-9

Özgün-Koca, S. A., & İlhan Şen, A. (2006). The beliefs and perceptions of pre-service teachers enrolled in a subject-area dominant teacher education program about "Effective Education". *Teaching and Teacher Education, 22*(7), 946–960. https://doi.org/10.1016/j.tate.2006.04.036

Potari, D., & Chapman, O. (Eds.). (2020*). International Handbook of Mathematics Teacher Education: Volume 1: Knowledge, beliefs, and identity in mathematics teaching and teaching development.* Brill.

Piwowar, V., Barth, V. L., Ophardt, D., & Thiel, F. (2018). Evidence-based scripted videos on handling student misbehavior: the development and evaluation of video cases for teacher education. *Professional Development in Education, 44*(3), 369–384. https://doi.org/10.1080/19415257.2017.1316299

Santagata, R., König, J., Scheiner, T., Nguyen, H., Adleff, A.-K., Yang, X., & Kaiser, G. (2021). Mathematics teacher learning to notice: A systematic review of studies of video-supported teacher education. *ZDM Mathematics Education, 53,* 119–134. https://doi.org/10.1007/s11858-020-01216-z

Santagata, R., & Yeh, C. (2016). The role of perception, interpretation, and decision making in the development of beginning teachers' competence. *ZDM Mathematics Education, 48*(1-2), 153–165.

Scheiner, T. (2021). Towards a more comprehensive model of teacher noticing. ZDM—Mathematics Education, 53(1), 85–94. https://doi.org/10.1007/s11858-020-01202-5

Scheiner, T. (2023). Shifting the ways prospective teachers frame and notice student mathematical thinking: From deficits to strengths. *Educational Studies in Mathematics,* 1–27. https://doi.org/10.1007/s10649-023-10235-y

Schlag, S., & Glock, S. (2019). Entwicklung von Wissen und selbsteingeschätztem Wissen zur Klassenführung während des Praxissemesters im Lehramtsstudium. *Unterrichtswissenschaft, 47*(2), 221–241. https://doi.org/10.1007/s42010-019-00037-8

Scholten, N., & Orschulik, A. (2022). Praxisdokumente zur Verknüpfung von Theorie und Praxis auf Basis der Professionellen Unterrichtswahrnehmung. *Herausforderungen Lehrer*innenbildung—Zeitschrift für Konzeption, Gestaltung und Diskussion, 5*(1), 179–195. https://doi.org/10.11576/HLZ-5233

Schubarth, W., Speck, K., Seidel, A., Gottmann, C., Kamm, C., & Krohn, M. (2012). "Das Praxissemester im Lehramt–ein Erfolgsmodell? Zur Wirksamkeit des Praxissemesters im Land Brandenburg," in *Studium nach Bologna: Praxisbezüge stärken?! Praktika als Brücke zwischen Hochschule und Arbeitsmarkt*, eds. W. Schubarth, K. Speck, A. Seidel, C. Gottmann, C. Kamm, & M. Krohn (Springer VS), 137–169. https://doi.org/10.1007/978-3-531-19122-5_8

Seidel, T., & Stürmer, K. (2014). Modeling and measuring the structure of professional vision in preservice teachers. *American Educational Research Journal, 51*(4), 739–771. https://doi.org/10.3102/0002831214531321

Seifert, A., & Schaper, N. (2018). "Die Veränderung von Selbstwirksamkeitserwartungen und der Berufswahlsicherheit im Praxissemester: Empirische Befunde zur Bedeutung von Lerngelegenheiten und berufsspezifischer Motivation der Lehramtsstudierenden," in *Learning to practice, learning to reflect? Ergebnisse aus der Längsschnittstudie LtP zur Nutzung und Wirkung des Praxissemesters in der Lehrerbildung*, eds. J. König, M. Rothland, & N. Schaper (Springer VS), 195–222.

Seifert, A., Schaper, N., & König, J. (2018). "Bildungswissenschaftliches Wissen und Kompetenzeinschätzungen von Studierenden im Praxissemester: Veränderungen und Zusammenhänge," in *Learning to practice, learning to reflect? Ergebnisse aus der Längsschnittstudie LtP zur Nutzung und Wirkung des Praxissemesters in der Lehrerbildung*, eds. J. König, M. Rothland, & N. Schaper (Springer VS), 325–347.

Shah, N., & Coles, J. A. (2020). Preparing Teachers to Notice Race in Classrooms: Contextualizing the Competencies of Preservice Teachers With Antiracist Inclinations. *Journal of Teacher Education, 71*(5), 584–599. https://doi.org/10.1177/0022487119900204

Sherin, M. G. (2007). The development of teachers' professional vision in video clubs. In R. Goldman, R. Pea, B. Barron, & S. Derry (Eds.), *Video Research in the Learning Sciences* (pp. 383–395). Erlbaum.

Sherin, M. G., Jacobs, V. R., & Philipp, R. A. (2011). Situating the study of teachers' noticing. In M. G. Sherin, V. R. Jacobs, & R. A. Philipp (Eds.), *Studies in mathematical thinking and learning. Mathematics teachers' noticing* (pp. 3–13). Routledge.

Stigler, J. W., & Miller, K. F. (2018). "Expertise and expert performance in teaching," in *The Cambridge handbook of expertise and expert performance*, eds. K. A. Ericsson, R. R. Hoffman, A. Kozbelt, & M. A. Williams (Cambridge University Press), 431–452. https://doi.org/10.1017/9781316480748.024

Stürmer, K., Seidel, T., & Schäfer, S. (2013). Changes in professional vision in the context of practice. *Gruppendynamik Und Organisationsberatung, 44*(3), 339–355. https://doi.org/10.1007/s11612-013-0216-0

Superfine, A. C., Fisher, A., Bragelman, J., & Amador, J. M. (2017). "Shifting perspectives on preservice teachers' noticing of children's mathematical thinking", in *Research in Mathematics Education. Teachers' noticing: Bridging and broadening perspectives, contexts, and frameworks*, eds. E. O. Schack, M. H. Fisher, & J. A. Wilhelm (Springer International Publishing), 409–426.

Terhart, E. (2021). "Teacher education in Germany: Historical development, status, reforms and challenges," in *Asia-Europe education dialogue. Quality in teacher education and professional development: Chinese and German perspectives*, eds. J. C.-K. Lee & T. Ehmke (Routledge Taylor & Francis), 44–56.

Thomas, J., Dueber, D., Fisher, M. H., Jong, C., & Schack, E. O. (2021). Professional noticing coherence: exploring relationships between component processes. *Mathematical Thinking and Learning*, 1–19. https://doi.org/10.1080/10986065.2021.1977086

Ulrich, I., Klingebiel, F., Bartels, A., Staab, R., Scherer, S., & Gröschner, A. (2020). "Wie wirkt das Praxissemester im Lehramtsstudium auf Studierende? Ein systematischer Review," *Praxissemester im Lehramtsstudium in Deutschland: Wirkungen auf*

Studierende, eds. in I. Ulrich & A. Gröschner (Springer Fachmedien), 1–66. https://doi.org/10.1007/978-3-658-24209-1_1

van Es, E. A. (2011). A framework for learning to notice student thinking. In M. G. Sherin, V. R. Jacobs, & R. A. Philipp (Eds.), *Studies in mathematical thinking and learning. Mathematics teachers' noticing* (pp. 134–151). Routledge.

van Es, E. A., & Sherin, M. G. (2002). Learning to notice: Scaffolding new teachers' interpretations of classroom interactions. *Journal of Technology and Teacher Education, 10*, 571–596.

Weber, K. E., Gold, B., Prilop, C. N., & Kleinknecht, M. (2018). Promoting pre-service teachers' professional vision of classroom management during practical school training: Effects of a structured online- and video-based self-reflection and feedback intervention. *Teaching and Teacher Education, 76*, 39–49. https://doi.org/10.1016/j.tate.2018.08.008

Weber, K. E., Prilop, C. N., Viehoff, S., Gold, B., & Kleinknecht, M. (2020). Fördert eine videobasierte Intervention im Praktikum die professionelle Wahrnehmung von Klassenführung?—Eine quantitativ-inhaltsanalytische Messung von Subprozessen professioneller Wahrnehmung. *Zeitschrift für Erziehungswissenschaft, 23*(2), 343–365. https://doi.org/10.1007/s11618-020-00939-9

Weinert, F. E. (2001). Concept of Competence: A conceptual clarification. In D. S. Rychen & L. H. Salganik (Eds.), *Defining and selecting key competencies* (pp. 45–65). Hogrefe & Huber Publishers.

Weyers, J., König, J., Rott, B., Greefrath, G., Vorhölter, K., & Kaiser, G. (2023). Mathematics teachers' professional noticing: Transfer of a video-based competence assessment instrument into teacher education for evaluation purposes. *Zeitschrift Für Erziehungswissenschaft*, 1–26. https://doi.org/10.1007/s11618-023-01159-7

Weyers, J., König, J., Santagata, R., Scheiner, T., & Kaiser, G. (2023). Measuring teachers' noticing: A scoping review of standardized instruments. *Teaching and Teacher Education, 122*. https://doi.org/10.1016/j.tate.2022.103970

Weyers, J., Kramer, C., Kaspar, K., & König, J. (2024). Measuring pre-service teachers' decision-making in classroom management: A video-based assessment approach. *Teaching and Teacher Education*, 138, 104426. https://doi.org/10.1016/j.tate.2023.104426

Yang, X., König, J., & Kaiser, G. (2021). Growth of professional noticing of mathematics teachers: A comparative study of Chinese teachers noticing with different teaching experience. *ZDM Mathematics Education, 53*, 29–42. https://doi.org/10.1007/s11858-020-01217-y

7 Discussion of Results and Outlook

In this thesis, I examined the teacher noticing skills, namely perception, interpretation, and decision-making, of pre-service and in-service secondary mathematics teachers using a video-based instrument to gain insight into the structure and development of teacher noticing, as well as factors that influence the construct. This research objective was addressed through three separate publications, each focusing on a specific part of the objective. After describing the publications and imprinting them in the previous chapters, I briefly summarize and discuss the results in this section. Finally, I provide an outlook for research on teacher noticing and some concluding remarks.

7.1 Summary and Discussion

The analytical conceptualization of teacher noticing with its three facets of perception, interpretation, and decision-making was empirically separable and proved to be superior to a single-factor model using a single scale, thus corresponding to a holistic understanding of noticing at the conceptual level (Bastian et al., 2022). Therefore, the results provided strong evidence for the distinctiveness of the three facets, and distinguishing between the facets of teacher noticing was shown to be important for gaining detailed insight into the construct. In addition, decision-making was confirmed as an essential part of teacher noticing, as evidenced by strong correlations with the other two facets (Bastian et al., 2022; Bastian et al., 2024). This may challenge conceptualizations that only include perceptual and/or interpretive facets (Seidel & Stürmer 2014; Star & Strickland, 2008).

A. Bastian, *Teacher Noticing of Pre-service and In-service Secondary Mathematics Teachers*, Perspektiven der Mathematikdidaktik,
https://doi.org/10.1007/978-3-658-45619-1_7

The findings of this thesis indicate complex, reciprocal relationships between the three facets in their development, as opposed to a purely linear learning process of first learning to perceive, then learning to interpret, and learning to make decisions. Interpretation appeared particularly important in the interaction structure, as it not only predicted its own development but also influenced the development of perception and decision-making (Bastian et al., 2024). The results even provided evidence for a causal relationship; that is, strong interpretation skills seemed to result in strong perception and decision-making skills (Bastian et al., 2024).

The development of teacher noticing over time and the influence of increasing teaching experience were investigated in all three publications of this thesis. The results conclusively demonstrated that a certain amount of teaching experience is necessary to develop skills in teacher noticing (Bastian et al., 2023; Bastian et al., 2022). In-service teachers outperformed pre-service teachers at the end of their university education, with large effect sizes and on nearly all cognitive demands of teacher noticing (Bastian et al., 2023; Bastian et al., 2022). Moreover, this thesis provided evidence of significant development over time in the course of teaching experience in a pretest-posttest design by examining the teaching internships of pre-service teachers (Bastian et al., 2024). However, the development of teacher noticing may not be linear growth, as saturation occurred (Bastian et al., 2022).

In a cross-sectional comparison between two groups of in-service teachers, only a few differences between the groups emerged, and they were in favor of early career teachers with less teaching experience (Bastian et al., 2022). The findings even suggested a decline in teacher noticing skills with increasing teacher experience from a certain point on (Bastian et al., 2022). The group of early career teachers outperformed experienced teachers in the areas of decision-making, mathematics pedagogical knowledge, and the handling of recent mathematics pedagogical and general pedagogical topics with small to moderate effect sizes (Bastian et al., 2023). These findings contradict results from other Western countries (Jacobs et al., 2010) and China (Yang, König, & Kaiser, 2021), which demonstrated more linear increases for the group of in-service teachers with increasing teaching experience, indicating cultural dependencies. Moreover, they may suggest that teachers first need teaching experience to develop teacher noticing skills, but at some point, teaching experience alone is no longer sufficient to progress further. Deliberate reflective practice may be necessary to further develop and reach the proficient or expert level (Ericsson et al., 1993). This is supported by the positive influences of reflective activities, in which teachers connect particular situations with theories of teaching and learning and their professional knowledge, on teacher noticing (Bastian et al., 2024). On the other hand, because the results of the better-performing early career teachers were gained cross-sectionally, and changes in the teacher education curriculum may have affected the groups, the findings may also be a sign of benefits from these reforms and changes. Nevertheless, it should be noted that experienced teachers appear to lack some skills, particularly in dealing with heterogeneity, competence-oriented teaching, and other topics of recent advances in mathematics pedagogy and general pedagogy (Bastian et al., 2023), which calls for professional development in these areas.

As noted above, teaching experience has been shown to have a significant impact on teacher noticing (Bastian et al., 2023; Bastian et al., 2022; Bastian et al., 2024). To gain further insight into which activities may specifically influence teacher noticing and which opportunities to learn may benefit teachers the most, I contributed to a study that longitudinally investigated the development of teacher noticing and related it to teaching practice activities in the context of teaching internships (Bastian et al., 2024). The results of the study showed that the amount of teaching over the course of the internship had no effect on teacher noticing by itself. However, the amount of time spent on lesson follow-up, the linking of the theories of university teacher education to specific situations, and the emotional support of a mentor had significant positive influences on facets of teacher noticing (Bastian et al., 2024). This, in turn, supports the importance of deliberate practice in which teachers are aware of what they are doing, have time to reflect and connect their practice to professional knowledge, and can share ideas and talk about their teaching with other teachers for the development of teacher noticing.

7.2 Prospects and Further Developments

With an extensive, rich, but also heterogeneous body of research on teacher noticing (König et al., 2022; Wei et al., 2023), the question arises as to how research on teacher noticing should develop and what future research objectives should be. Based on the findings of this thesis, there is a need for more professional development for in-service teachers, as well as deliberate practice opportunities for pre-service and in-service teachers. This calls for the development of these programs, especially in current issues in mathematics pedagogy and general pedagogy, such as dealing with heterogeneity and competence-oriented teaching. Moreover, more high-quality measurement instruments are needed to investigate teacher noticing skills in more detail and in specific domains, as development in and characteristics of teacher noticing may be domain specific (Bastian et al., 2023; Berliner, 2001; Weyers, König, Santagata, et al., 2023). A variety of methods to study teacher noticing may be useful for improving knowledge and gaining new insight, including the established video-based approach as well as advances with new technologies, such as 360° video (Kosko et al., 2021), eye tracking (Grub et al., 2022; Li et al., 2023), and the experimental approach of heart rate variation tracking (Kosko & Ferdig, 2023).

Regarding the conceptualization and frameworks of teacher noticing, the heterogeneous definitions and approaches based on different theoretical perspectives challenge the research community, since this heterogeneity hinders comparability (König et al., 2022). Therefore, researchers should explicitly describe their framework, place it in the current discussion, and link their terminology to other established conceptualizations to facilitate the interpretation of research results. Furthermore, the discussion could benefit from a shared understanding of the basic facets of teacher noticing and common terminology. Overview articles and literature reviews have taken the first steps in this direction (Amador et al., 2021; Dindyal et al., 2021; König et al., 2022; Scheiner & Kaiser, 2023; M. G. Sherin, Jacobs, & Philipp, 2011b). In this context, the findings of this thesis support the need for a decision-oriented facet of teacher noticing. As the current frameworks are strongly influenced by researchers from Western countries (König et al., 2022; Scheiner & Kaiser, 2023; Wei et al., 2023), frameworks from other cultures, such as the lesson-study influenced FOCUS framework (Choy, 2015) or the focused framework from China (Zhang et al., 2023), and culturally sensitive frameworks should also be included in the discussion and consideration of an overall basic framework. Other facets such as the enactment of teacher noticing (Thomas et al., 2020) could also be included.

To address the inconclusive findings regarding the development of teacher noticing with increasing teaching experience (Bastian et al., 2022; Jacobs et al., 2010; Yang, König, & Kaiser, 2021), more longitudinal, quantitative studies are needed, as most results to date are based on cross-sectional analyses. Finally, as teacher noticing is part of teachers' professional competence and thus of the complex interaction patterns of teaching and learning, we need more studies that explore these relationships and take into account individual and contextual aspects to better understand teacher noticing and to find improved means to facilitate its development through teacher education and professional development.

Appendix

List of the Publications in the Context of This Dissertation

Bastian, A., Hoth, J., Larrain Jory, M., & Yang, X. (2023). New directions in research on teacher noticing. In M. Ayalon, B. Koichu, R. Leikin, L. Rubel, & M. Tabach (Eds.), *Proceedings of the 46th conference of the International Group for the Psychology of Mathematics Education* (Vol. 1, pp. 118–122). PME.

Bastian, A., Kaiser, G., Meyer, D., & König, J. (2024). The link between expertise, the cognitive demands of teacher noticing and, experience in teaching mathematics in secondary schools. International Journal of Science and Mathematics Education, 22, 257-282. https://doi.org/10.1007/s10763-023-10374-x

Bastian, A., Kaiser, G., Meyer, D., Schwarz, B., & König, J. (2022). Teacher noticing and its growth toward expertise: An expert–novice comparison with pre-service and in-service secondary mathematics teachers. *Educational Studies in Mathematics, 110*(2), 205–232. https://doi.org/10.1007/s10649-021-10128-y

Bastian, A., König, J., Weyers, J., Siller, H.-S., & Kaiser, G. (2024). Effects of teaching internships on preservice teachers' noticing in secondary mathematics education. *Frontiers in Education, 9,* 1-15. https://doi.org/10.3389/feduc.2024.1360315

Bastian, A., Rehbock, D., Schmiedel, H., & Weigand, H. (2022). Professionelle Unterrichtswahrnehmung – Konstruktverständnis von (zukünftigen) Lehrkräften. In N. Buchholtz, B. Schwarz, & K. Vorhölter (Eds.), *Initiationen mathematikdidaktischer Forschung* (pp. 229–246). Springer Spektrum. https://doi.org/10.1007/978-3-658-36766-4_12

A. Bastian, *Teacher Noticing of Pre-service and In-service Secondary Mathematics Teachers*, Perspektiven der Mathematikdidaktik,
https://doi.org/10.1007/978-3-658-45619-1

References

Adams, R. J., Wu, M., Macaskill, G., Haldane, S., Sun, X. X., Cloney, D., & Berezner, A. (1997–2023). *ConQuest* (Version 5.28) [Computer software]. Australian Council for Educational Research. www.acer.org/gb/conquest

Amador, J. M., Bragelman, J., & Superfine, A. C. (2021). Prospective teachers' noticing: A literature review of methodological approaches to support and analyze noticing. *Teaching and Teacher Education, 99*, 1–16. https://doi.org/10.1016/j tate.2020.103256

American Educational Research Association, American Psychological Association, National Council on Measurement in Education (Ed.). (2014). *Standards for educational and psychological testing.* American Educational Research Association.

Anderson, J., & Taner, G. (2023). Building the expert teacher prototype: A metasummary of teacher expertise studies in primary and secondary education. *Educational Research Review, 38*, 100485. https://doi.org/10.1016/j.edurev.2022.100485

Bastian, A., Hoth, J., Larrain Jory, M., & Yang, X. (2023). New directions in research on teacher noticing. In M. Ayalon, B. Koichu, R. Leikin, L. Rubel, & M. Tabach (Eds.), *Proceedings of the 46th conference of the International Group for the Psychology of Mathematics Education* (Vol. 1, pp. 118–122). PME.

Bastian, A., Kaiser, G., Meyer, D., & König, J. (2024). The link between expertise, the cognitive demands of teacher noticing and, experience in teaching mathematics in secondary schools. International Journal of Science and Mathematics Education, 22, 257-282. https://doi.org/10.1007/s10763-023-10374-x

Bastian, A., Kaiser, G., Meyer, D., Schwarz, B., & König, J. (2022). Teacher noticing and its growth toward expertise: An expert–novice comparison with pre-service and in-service secondary mathematics teachers. *Educational Studies in Mathematics, 110*(2), 205–232. https://doi.org/10.1007/s10649-021-10128-y

A. Bastian, *Teacher Noticing of Pre-service and In-service Secondary Mathematics Teachers*, Perspektiven der Mathematikdidaktik,
https://doi.org/10.1007/978-3-658-45619-1

Bastian, A., König, J., Weyers, J., Siller, H.-S., & Kaiser, G. (2024). Effects of teaching internships on preservice teachers' noticing in secondary mathematics education. *Frontiers in Education, 9,* 1-15. https://doi.org/10.3389/feduc.2024.1360315

Bastian, A., Rehbock, D., Schmiedel, H., & Weigand, H. (2022). Professionelle Unterrichtswahrnehmung – Konstruktverständnis von (zukünftigen) Lehrkräften [Teacher noticing: Construct understanding of pre-service and in-service teachers]. In N. Buchholtz, B. Schwarz, & K. Vorhölter (Eds.), *Initiationen mathematikdidaktischer Forschung* (pp. 229–246). Springer Spektrum. https://doi.org/10.1007/978-3-658-36766-4_12

Baumert, J., & Kunter, M. (2006). Stichwort: Professionelle Kompetenz von Lehrkräften [Keyword: Professional competence of teachers]. *Zeitschrift für Erziehungswissenschaft, 9*(4), 469–520. https://doi.org/10.1007/s11618-006-0165-2

Baumert, J., & Kunter, M. (2013). The COACTIV model of teachers' professional competence. In M. Kunter, J. Baumert, W. Blum, U. Klusmann, S. Krauss, & M. Neubrand (Eds.), *Cognitive Activation in the Mathematics Classroom and Professional Competence of Teachers* (25-48). Springer New York.

Berliner, D. C. (1986). In pursuit of the expert pedagogue. *Educational Researcher, 15*(7), 5. https://doi.org/10.2307/1175505

Berliner, D. C. (1988). *The development of expertise in pedagogy.* American Association of Colleges for Teacher Education, Washington, D.C.

Berliner, D. C. (2001). Learning about and learning from expert teachers. *International Journal of Educational Research, 35*(5), 463–482. https://doi.org/10.1016/S0883-0355(02)00004-6

Berliner, D. C. (2004). Expert teachers: Their characteristics, development and accomplishments. In R. Batllori i Obiols, A. E Gomez Martinez, M. Oller i Freixa, & J. Pages i Blanch (Eds.), *De la teoria…a l'aula: Formacio del professorat ensenyament de las ciències socials* (pp. 13–28). Departament de Didàctica de la Llengua de la Literatura I de les Ciències Socials, Universitat Autònoma de Barcelona.

Blömeke, S., Bremerich-Vos, A., Haudeck, H., Kaiser, G., Nold, G., Schwippert, K., & Willenberg, H. (Eds.). (2011). *Kompetenzen von Lehramtsstudierenden in gering strukturierten Domänen: Erste Ergebnisse aus TEDS-LT* [Competencies of pre-service teachers in low-structured domains: Initial results from TEDS-LT]. Waxmann. http://www.content-select.com/index.php?id=bib_view&ean=9783830975106

Blömeke, S., Bremerich-Vos, A., & Kaiser, G. (Eds.). (2013). *Professionelle Kompetenzen im Studienverlauf: Weitere Ergebnisse zur Deutsch-, Englisch- und Mathematiklehrerausbildung aus TEDS-LT* [Professional competence in the course of study: Further findings on German, Englisch and mathematics teacher education form TEDS-LT]. Waxmann. http://ebooks.ciando.com/book/index.cfm/bok_id/1184754

Blömeke, S., Busse, A., Kaiser, G., König, J., & Suhl, U. (2016). The relation between content-specific and general teacher knowledge and skills. *Teaching and Teacher Education, 56*, 35–46. https://doi.org/10.1016/j.tate.2016.02.003

Blömeke, S., Gustafsson, J. E., & Shavelson, R. J. (2015). Beyond dichotomies: Competence viewed as a continuum. *Zeitschrift Für Psychologie / Journal of Psychology, 223*(1), 3–13. https://doi.org/10.1027/2151-2604/a000194

Blömeke, S., Houang, R. T., & Suhl, U. (2014). Diagnosing teacher knowledge by applying multidimensional item response theory and multiple-group models. In S. Blömeke, F.-J. Hsieh, G. Kaiser, & W. H. Schmidt (Eds.), *Advances in mathematics education: Vol.*

6. International perspectives on teacher knowledge, beliefs and opportunities to learn: TEDS-M Results (pp. 483–501). Springer Spektrum.
Blömeke, S., Hsieh, F.-J., Kaiser, G., & Schmidt, W. H. (Eds.). (2014). Advances in mathematics education: Vol. 6. International perspectives on teacher knowledge, beliefs and opportunities to learn: TEDS-M Results. Springer Spektrum. https://doi.org/10.1007/978-94-007-6437-8
Blömeke, S., Jentsch, A., Ross, N., Kaiser, G., & König, J. (2022). Opening up the black box: Teacher competence, instructional quality, and students' learning progress. *Learning and Instruction, 79*, 1–11. https://doi.org/10.1016/j.learninstruc.2022.101600
Blömeke, S., & Kaiser, G. (2014). Theoretical framework, study design and main results of TEDS-M. In S. Blömeke, F.-J. Hsieh, G. Kaiser, & W. H. Schmidt (Eds.), *Advances in mathematics education: Vol. 6. International perspectives on teacher knowledge, beliefs and opportunities to learn: TEDS-M Results* (pp. 19–47). Springer Spektrum.
Blömeke, S., & Kaiser, G. (2017). Understanding the development of teachers' professional competencies as personally, sitationally and socially determined. In D. J. Clandini & J. Husu (Eds.), *International handbook on research on teacher education* (pp. 783–802). SAGE Publications Inc.
Blömeke, S., Kaiser, G., & Lehmann, R. (Eds.). (2010a). *TEDS-M 2008: Professionelle Kompetenz und Lerngelegenheiten angehender Mathematiklehrkräfte für die Sekundarstufe I im internationalen Vergleich* [TEDS-M 2008: Professional competence and opportunities to learn of pre-service secondary mathematics teachers in international comparison]. Waxmann.
Blömeke, S., Kaiser, G., & Lehmann, R. (Eds.). (2010b). *TEDS-M 2008: Professionelle Kompetenz und Lerngelegenheiten angehender Primarstufenlehrkräfte im internationalen Vergleich* [TEDS-M 2008: Professional competence and opportunities to learn of pre-service primary mathematics teachers in international comparison]. Waxmann.
Blömeke, S., & König, J. (2010a). Messung des pädagogischen Wissens: Theoretischer Rahmen und Teststruktur [Measurement of general pedagogical knowledge: Theoretical framework and test structure]. In S. Blömeke, G. Kaiser, & R. Lehmann (Eds.), *TEDS-M 2008: Professionelle Kompetenz und Lerngelegenheiten angehender Mathematiklehrkräfte für die Sekundarstufe I im internationalen Vergleich* (pp. 239–264). Waxmann.
Blömeke, S., & König, J. (2010b). Pädagogisches Wissen angehender Mathematiklehrkräfte für die Sekundarstufe I im internationalen Vergleich [General pedagogical knowledge of pre-service secondary mathematics teachers in international comparison]. In S. Blömeke, G. Kaiser, & R. Lehmann (Eds.), *TEDS-M 2008: Professionelle Kompetenz und Lerngelegenheiten angehender Mathematiklehrkräfte für die Sekundarstufe I im internationalen Vergleich* (pp. 265–278). Waxmann.
Blömeke, S., König, J., Busse, A., Suhl, U., Benthien, J., Döhrmann, M., & Kaiser, G. (2014). Von der Lehrerausbildung in den Beruf - Fachbezogenes Wissen als Voraussetzung für Wahrnehmung, Interpretation und Handeln im Unterricht [From teacher education to teaching practice: Subject-related knowledge as a prerequisite for perception, interpretation and decision-making in the classroom]. *Zeitschrift Für Erziehungswissenschaft, 17*(3), 509–542. https://doi.org/10.1007/s11618-014-0564-8
Blömeke, S., König, J., Suhl, U., Hoth, J., & Döhrmann, M. (2015). Wie situationsbezogen ist die Kompetenz von Lehrkräften? Zur Generalisierbarkeit der Ergebnisse von

videobasierten Performanztests [How situation-specific is teacher competence? On the generalizability of the results of a video-based performance test]. *Zeitschrift Für Pädagogik*, *61*(3), 310–327.

Blum, W. (2019). Unterrichtsqualität aus fachdidaktischer Perspektive: Beispiele aus der Mathematik [Instructional quality from a perspective of subjects' pedagogy: Example from mathematics]. In U. Steffens & R. Messner (Eds.), *Beiträge zu Schulentwicklung: Grundlagen der Qualität von Schule 3. Unterrichtsqualität - Konzepte und Bilanzen gelingenden Lehrens und Lernens: Konzepte und Bilanzen gelingenden Lehrens und Lernens* (pp. 183–200). Waxmann.

Bond, T. G., Yan, Z., & Heene, M. (2021). *Applying the Rasch model: Fundamental measurement in the human sciences* (Fourth edition). Routledge Taylor & Francis Group.

Bromme, R. (1992). *Der Lehrer als Experte: Zur Psychologie des professionellen Wissens* [The teacher as an expert: About the psychology of professional knowledge]. Huber.

Buchholtz, N., & Kaiser, G. (2013). Improving mathematics teacher education in Germany: Empirical results from a longitudinal evaluation of innovative programs. *International Journal of Science and Mathematics Education*, *11*(4), 949–977. https://doi.org/10.1007/s10763-013-9427-7

Carter, K., Cushing, K. S., Sabers, D. S., Stein, P., & Berliner, D. C. (1988). Expert-novice differences in perceiving and processing visual classroom information. *Journal of Teacher Education*, *39*(3), 25–31. https://doi.org/10.1177/002248718803900306

Caspari-Sadeghi, S., & König, J. (2018). On the adequacy of expert teachers: From practical convenience to psychological reality. *International Journal of Higher Education*, *7*(5), 1–19. https://doi.org/10.5430/ijhe.v7n5p1

Chan, K. K. H., Xu, L., Cooper, R., Berry, A., & van Driel, J. H. (2021). Teacher noticing in science education: do you see what I see? *Studies in Science Education*, *57*(1), 1–44. https://doi.org/10.1080/03057267.2020.1755803

Choy, B. H. (2015). *The FOCUS framework - Snapshots of mathematicss teacher noticing.* Doctoral thesis. The University of Auckland.

Cohen, J. (1960). A coefficient of agreement for nominal scales. *Educational and Psychological Measurement*, *20*(1), 37–46. https://doi.org/10.1177/001316446002000104

Cohen, J. (1988). *Statistical power analysis for the behavioral sciences* (2.th ed.). Lawrence Erlbaum Associates.

Collins, H., & Evans, R. (2018). A sociological/philosophical perspective on expertise: The acquisition of expertise through socialization. In K. A. Ericsson, R. R. Hoffman, A. Kozbelt, & M. A. Williams (Eds.), *The Cambridge handbook of expertise and expert performance* (pp. 21–32). Cambridge University Press.

Copur-Gencturk, Y., & Tolar, T. (2022). Mathematics teaching expertise: A study of the dimensionality of content knowledge, pedagogical content knowledge, and content-specific noticing skills. *Teaching and Teacher Education*, *114*, 103696. https://doi.org/10.1016/j.tate.2022.103696

Cortina, K. S., & Thames, M. H. (2013). Teacher Education in Germany. In M. Kunter, J. Baumert, W. Blum, U. Klusmann, S. Krauss, & M. Neubrand (Eds.), *Cognitive Activation in the Mathematics Classroom and Professional Competence of Teachers* (pp. 49–62). Springer New York. https://doi.org/10.1007/978-1-4614-5149-5_3

Cramer, C., König, J., Rothland, M., & Blömeke, S. (Eds.). (2020). *Handbuch Lehrerinnen- und Lehrerbildung.* Verlag Julius Klinkhardt.

Dall'Alba, G. (2018). Reframing expertise and its development: A lifeworld perspective. In K. A. Ericsson, R. R. Hoffman, A. Kozbelt, & M. A. Williams (Eds.), *The Cambridge handbook of expertise and expert performance* (pp. 33–39). Cambridge University Press.

Depaepe, F., Verschaffel, L., & Kelchtermans, G. (2013). Pedagogical content knowledge: A systematic review of the way in which the concept has pervaded mathematics educational research. *Teaching and Teacher Education, 34*, 12–25. https://doi.org/10.1016/j.tate.2013.03.001

Dindyal, J., Schack, E. O., Choy, B. H., & Sherin, M. G. (2021). Exploring the terrains of mathematics teacher noticing. *ZDM – Mathematics Education, 53*(1), 1–16. https://doi.org/10.1007/s11858-021-01249-y

Döhrmann, M., Kaiser, G., & Blömeke, S. (2010). Messung des mathematischen und mathematikdidaktischen Wissens: Theoretischer Rahmen und Teststruktur [Measurement of mathematics and mathematics pedagogical knowledge: Theoretical framework and test structure]. In S. Blömeke, G. Kaiser, & R. Lehmann (Eds.), *TEDS-M 2008: Professionelle Kompetenz und Lerngelegenheiten angehender Mathematiklehrkräfte für die Sekundarstufe I im internationalen Vergleich* (pp. 169–196). Waxmann.

Döhrmann, M., Kaiser, G., & Blömeke, S. (2012). The conceptualisation of mathematics competencies in the international teacher education study TEDS-M. *ZDM – Mathematics Education, 44*(3), 325–340. https://doi.org/10.1007/s11858-012-0432-z

Doyle, W. (1986). Classroom organization and management. In M. C. Wittrock (Ed.), *Handbook of research on teaching* (pp. 392–431). Macmillan.

Drahmann, M. (2020). Teacher Education in Germany: A Holistic View of Structure, Curriculum, Development and Challenges. In K. Pushpanadham (Ed.), *Springer eBook Collection. Teacher Education in the Global Era: Perspectives and Practices* (1st ed. 2020, pp. 13–31). Springer Singapore; Imprint Springer. https://doi.org/10.1007/978-981-15-4008-0_2

Dreher, A., & Kuntze, S. (2015). Teachers' professional knowledge and noticing: The case of multiple representations in the mathematics classroom. *Educational Studies in Mathematics, 88*(1), 89–114. https://doi.org/10.1007/s10649-014-9577-8

Dreyfus, H. L., & Dreyfus, S. E. (1986). Mind over machine: The power of human intuition and expertise in the era of the computer. Basil Blackwell Ltd.

Eckhardt, T. (2021). The education system in the Federal Republic of Germany 2018/2019: A description of the responsibilities, structures and developments in education policy for the exchange of information in Europe. https://www.kmk.org/fileadmin/Dateien/pdf/Eurydice/Bildungswesen-engl-pdfs/dossier_en_ebook.pdf

Erickson, F. (2011). On noticing teacher noticing. In M. G. Sherin, V. R. Jacobs, & R. A. Philipp (Eds.), *Studies in mathematical thinking and learning. Mathematics teacher noticing* (pp. 17–34). Routledge.

Ericsson, K. A. (2018). An introduction to the second edition of The Cambridge Handbook of Expertise and Expert Performance: Its development, organization, and content. In K. A. Ericsson, R. R. Hoffman, A. Kozbelt, & M. A. Williams (Eds.), *The Cambridge handbook of expertise and expert performance* (pp. 3–20). Cambridge University Press.

Ericsson, K. A., Krampe, R. T., & Tesch-Römer, C. (1993). The role of deliberate practice in the acquisition of expert performance, *100*(3), 363–406. https://doi.org/10.1037/0033-295X.100.3.363

European Agency for Development in Special Needs Education. (2012). *Teacher education for inclusion: Profile of inclusive teachers.* https://www.european-agency.org/sites/default/files/profile_of_inclusive_teachers_en.pdf

Felske, C., König, J., Kaiser, G., Klemenz, S., Ross, N., & Blömeke, S. (2020). Pädagogisches Wissen von berufstätigen Mathematiklehrkräften [General pedagogical knowlege of in-service mathematics teachers]. *Diagnostica, 66*(2), 110–122. https://doi.org/10.1026/0012-1924/a000250

Gegenfurtner, A. (2020). *Professional vision and visual expertise.* Universität Regensburg.

Gibson, J. J. (1979). *The ecological approach to visual perception.* Houghton Mifflin.

Gold, B., & Holodynski, M. (2017). Using digital video to measure the professional vision of elementary classroom management: Test validation and methodological challenges. *Computers & Education, 107*, 13–30. https://doi.org/10.1016/j.compedu.2016.12.012

Goodwin, C. (1994). Professional vision. *American Anthropologist, 96*(3), 606–633.

Grub, A.-S., Biermann, A [Anthe], & Brünken, R. (2020). Process-based measurement of professional vision of (prospective) teachers in the field of classroom management: A systematic review. *Journal for Educatinoal Research Online, 12*(3), 75–102.

Grub, A.-S., Biermann, A [Antje], Lewalter, D., & Brünken, R. (2022). Professional Vision and the Compensatory Effect of a Minimal Instructional Intervention: A Quasi-Experimental Eye-Tracking Study With Novice and Expert Teachers. *Frontiers in Education, 7*, Article 890690. https://doi.org/10.3389/feduc.2022.890690

Hattie, J. (2009). *Visible learning: A synthesis of over 800 meta-analyses relating to achievement.* Routledge.

Helmke, A., & Schrader, F. (2022). Angebots-Nutzungs-Modell der Wirkfaktoren akademischer Leistungen. In M. A. Wirtz (Ed.), *Dorsch - Lexikon der Psychologie.* Hogrefe.

Helsper, W. (2007). Eine Antwort auf Jürgen Baumerts und Mareike Kunters Kritik am strukturtheoretischen Professionsansatz [A response to Jürgen Baumert's and Mareike Kunter's critique of the structural-theoretical approach to profession]. *Zeitschrift für Erziehungswissenschaft, 10*(4), 567–579. https://doi.org/10.1007/s11618-007-0064-1

Hill, H. C., Ball, D. L., Blunk, M., Goffney, I. M., & Rowan, B. (2007). Validating the Ecological Assumption: The Relationship of Measure Scores to Classroom Teaching and Student Learning. *Measurement: Interdisciplinary Research and Perspectives, 5*(2–3), 107–118. https://doi.org/10.1080/15366360701487138

Hoth, J., Schwarz, B., Kaiser, G., Busse, A., König, J., & Blömeke, S. (2016). Uncovering predictors of disagreement: ensuring the quality of expert ratings. *ZDM Mathematics Education, 48*(1-2), 83–95. https://doi.org/10.1007/s11858-016-0758-z

Hughes, R., & Huby, M. (2002). The application of vignettes in social and nursing research. *Journal of Advanced Nursing, 37*(4), 382–386. https://doi.org/10.1046/j.1365-2648.2002.02100.x

Jacobs, V. R., Empson, S. B., Jessup, N. A., Dunning, A., Pynes, D., Krause, G., & Franke, T. M. (2022). Profiles of teachers' expertise in professional noticing of children's mathematical thinking. *Journal of Mathematics Teacher Education*, 1–30. https://doi.org/10.1007/s10857-022-09558-z

Jacobs, V. R., Lamb, L. L. C., & Philipp, R. A. (2010). Professional noticing of children's mathematical thinking. *Journal for Research in Mathematics Education, 41*(2), 169–202. www.jstor.org/stable/20720130

Jacobs, V. R., & Spangler, D. A. (2017). Research on core practices in K-12 mathematics teaching. In J. Cai (Ed.), *Compendium for research in mathematics education* (pp. 766–792). National Council of Teachers of Mathematics.

Jentsch, A., Schlesinger, L., Heinrichs, H., Kaiser, G., König, J., & Blömeke, S. (2021). Erfassung der fachspezifischen Qualität von Mathematikunterricht: Faktorenstruktur und Zusammenhänge zur professionellen Kompetenz von Mathematiklehrpersonen [Assessment of subject specific quality of mathematics teaching: Factor structure and correlations to the professional competence of teachers]. *Journal für Mathematik-Didaktik*, *42*(1), 97–121. https://doi.org/10.1007/s13138-020-00168-x

Jong, C., Schack, E. O., Fisher, M. H., Thomas, J., & Dueber, D. (2021). What role does professional noticing play? Examining connections with affect and mathematical knowledge for teaching among preservice teachers. *ZDM – Mathematics Education*, *53*(1), 151–164. https://doi.org/10.1007/s11858-020-01210-5

Kaiser, G., Blömeke, S., König, J., Busse, A., Döhrmann, M., & Hoth, J. (2017). Professional competencies of (prospective) mathematics teachers—cognitive versus situated approaches. *Educational Studies in Mathematics*, *94*(2), 1–22. https://doi.org/10.1007/s10649-016-9713-8

Kaiser, G., Busse, A., Hoth, J., König, J., & Blömeke, S. (2015). About the complexities of video-based assessments: Theoretical and methodological approaches to overcoming shortcomings of research on teachers' competence. *International Journal of Science and Mathematics Education*, *13*(2), 369–387. https://doi.org/10.1007/s10763-015-9616-7

Kaiser, G., & König, J. (2019). Competence measurement in (mathematics) teacher education and beyond: Implications for policy. *Higher Education Policy*(32), 597–615. https://doi.org/10.1057/s41307-019-00139-z

Kaiser, G., Scheiner, T., Ayalon, M., Kosko, K. W., Kersting, N. B., Fernandez, C., Superfine, A. C., Walkoe, J., Bastian, A., Hoth, J., Larrain Jory, M., Yang, X., & Choy, B. H. (2023). Innovative research approaches to mathematics teacher noticing. In M. Ayalon, B. Koichu, R. Leikin, L. Rubel, & M. Tabach (Eds.), *Proceedings of the 46th conference of the International Group for the Psychology of Mathematics Education* (pp. 103–133). PME.

Kane, M. T. (2013). Validating the Interpretations and Uses of Test Scores. *Journal of Educational Measurement*, *50*(1), 1–73. https://doi.org/10.1111/jedm.12000

Kelava, A., & Moosbrugger, H. (2011). Einführung in die Item-Response-Theorie (IRT) [Introduction into Item Response Theory (IRT)]. In H. Moosbrugger & A. Kelava (Eds.), *Springer-Lehrbuch. Testtheorie und Fragebogenkonstruktion* (2. Aufl. 2012, pp. 369–410). Springer Berlin Heidelberg.

Keppens, K., Consuegra, E., Goossens, M., Maeyer, S. de, & Vanderlinde, R. (2019). Measuring pre-service teachers' professional vision of inclusive classrooms: A video-based comparative judgement instrument. *Teaching and Teacher Education*, *78*, 1–14. https://doi.org/10.1016/j.tate.2018.10.007

Klieme, E., & Hartig, J. (2008). Kompetenzkonzepte in den Sozialwissenschaften und im erziehungswissenschaftlichen Diskurs [Competence concepts in the social and educational sciences]. In M. Prenzel, I. Gogolin, & H.-H. Krüger (Eds.), *Kompetenzdiagnostik: Zeitschrift für Erziehungswissenschaft* (pp. 11–29). VS Verlag für Sozialwissenschaften. https://doi.org/10.1007/978-3-531-90865-6_2

Klieme, E., Schümer, G., & Knoll, S. (2001). Mathematikunterricht in der Sekundarstufe I: „Aufgabenkultur“ und Unterrichtsgestaltung [Lower secondary mathematics teaching: „Task culture“ and lesson design]. In Bundesministerium für Bildung und Forschung (Ed.), *TIMSS–Impulse für Schule und Unterricht, Forschungsbefunde, Reforminitiativen, Praxisberichte und Video- Dokument* (pp. 43–58).

Koeppen, K., Hartig, J., Klieme, E., & Leutner, D. (2008). Current issues in competence modeling and assessment. *Zeitschrift Für Psychologie / Journal of Psychology, 216*(2), 61–73. https://doi.org/10.1027/0044-3409.216.2.61

Köller, O., Hasselhorn, M., Hesse, F. W., Maaz, K., Schrader, J., Solga, H., Spieß, C. K., & Zimmer, K. (Eds.). (2019). *Das Bildungswesen in Deutschland: Bestand und Potenziale* [The education system of Germany: Status survey and potentials]. Verlag Julius Klinkhardt. https://doi.org/10.36198/9783838547855

König, J. (2019). Empirische Befunde zu Effekten von Praxisphasen in der Lehrerausbildung [Empirical findings on effects of teaching internship in teacher education]. In N. McElvany, F. Schwabe, W. Bos, & H. G. Holtappels (Eds.), *IFS-Bildungsdialoge: Band 3. Lehrerbildung - Potentiale und Herausforderungen in den drei Phasen* (pp. 29–52). Waxmann.

König, J. (2020). Kompetenzorientierter Ansatz in der Lehrerinnen- und Lehrerbildung [The competence-oriented approach in teacher education]. In C. Cramer, J. König, M. Rothland, & S. Blömeke (Eds.), *utb Professionsforschung. Handbuch Lehrerinnen- und Lehrerbildung* (pp. 163–171).

König, J., Blömeke, S., Jentsch, A., Schlesinger, L., Felske née Nehls, C., Musekamp, F., & Kaiser, G. (2021). The links between pedagogical competence, instructional quality, and mathematics achievement in the lower secondary classroom. *Educational Studies in Mathematics, 107*(1), 189–212. https://doi.org/10.1007/s10649-020-10021-0

König, J., Blömeke, S., Klein, P., Suhl, U., Busse, A., & Kaiser, G. (2014). Is teachers' general pedagogical knowledge a premise for noticing and interpreting classroom situations? A video-based assessment approach. *Teaching and Teacher Education, 38*, 76–88. https://doi.org/10.1016/j.tate.2013.11.004

König, J., Doll, J., Buchholtz, N., Förster, S., Kaspar, K., Rühl, A.-M., Strauß, S., Bremerich-Vos, A., Fladung, I., & Kaiser, G. (2018). Pädagogisches Wissen versus fachdidaktisches Wissen? [General pedagogical knowledge versus subject specific pedagogical knowledge?]. *Zeitschrift für Erziehungswissenschaft, 21*(3), 1–38. https://doi.org/10.1007/s11618-017-0765-z

König, J., Santagata, R., Scheiner, T., Adleff, A.-K., Yang, X., & Kaiser, G. (2022). Teacher noticing: A systematic literature review of conceptualizations, research designs, and findings on learning to notice. *Educational Research Review, 36*. https://doi.org/10.1016/j.edurev.2022.100453

Kosko, K. W., & Ferdig, R. E. (2023). Relationships between prospective teachers' heart rate variation noticing of children's mathematics. In M. Ayalon, B. Koichu, R. Leikin, L. Rubel, & M. Tabach (Eds.), *Proceedings of the 46th conference of the International Group for the Psychology of Mathematics Education* (pp. 203–210). PME.

Kosko, K. W., Ferdig, R. E., & Zolfaghari, M. (2021). Preservice teachers' professional noticing when viewing standard and 360 video. *Journal of Teacher Education, 72*(3), 284–297. https://doi.org/10.1177/0022487120939544

Kounin, J. S. (1970). *Discipline and group management in classrooms*. Holt, Rinehart and Winston.

Kramer, C., König, J., Strauß, S., & Kaspar, K. (2020). Classroom videos or transcripts? A quasi-experimental study to assess the effects of media-based learning on pre-service teachers' situation-specific skills of classroom management. *International Journal of Educational Research, 103*, 101624. https://doi.org/10.1016/j.ijer.2020.101624

Krauss, S. (2020). Expertise-Paradigma in der Lehrerinnen- und Lehrerbildung [The expertise paradigm in teacher education]. In C. Cramer, J. König, M. Rothland, & S. Blömeke (Eds.), *utb Professionsforschung. Handbuch Lehrerinnen- und Lehrerbildung* (pp. 154–162).

Krauss, S., Bruckmaier, G., Lindl, A., Hilbert, S., Binder, K., Steib, N., & Blum, W. (2020). Competence as a continuum in the COACTIV study: The "cascade model". *ZDM Mathematics Education, 52*(2), 311–327. https://doi.org/10.1007/s11858-020-01151-z

Krauss, S., Brunner, M., Kunter, M., Baumert, J., Blum, W., Neubrand, M., & Jordan, A. (2008). Pedagogical content knowledge and content knowledge of secondary mathematics teachers. *Journal of Educational Psychology, 100*(3), 716–725. https://doi.org/10.1037/0022-0663.100.3.716

Krull, E., Oras, K., & Sisask, S. (2007). Differences in teachers' comments on classroom events as indicators of their professional development. *Teaching and Teacher Education, 23*(7), 1038–1050. https://doi.org/10.1016/j.tate.2006.02.001

Kunter, M., Baumert, J., Blum, W., Klusmann, U., Krauss, S., & Neubrand, M. (Eds.). (2013). *Cognitive Activation in the Mathematics Classroom and Professional Competence of Teachers*. Springer New York.

Lachner, A., Jarodzka, H., & Nückles, M. (2016). What makes an expert teacher? Investigating teachers' professional vision and discourse abilities. *Instructional Science, 44*(3), 197–203. https://doi.org/10.1007/s11251-016-9376-y

Leuders, T., Dörfler, T., Leuders, J., & Philipp, K. (2018). Diagnostic competence of mathematics teachers: Unpacking a complex construct. In T. Leuders, K. Philipp, & J. Leuders (Eds.), *Mathematics Teacher Education: volume 11. Diagnostic competence of mathematics teachers: Unpacking a complex construct in teacher education and teacher practice*. Springer.

Li, R., Cao, Y., Tang, H., & Kaiser, G. (2023). Teachers' Scaffolding Behavior and Visual Perception During Cooperative Learning. *International Journal of Science and Mathematics Education*, 1–20. https://doi.org/10.1007/s10763-023-10379-6

Lipowsky, F. (2006). Auf den Lehrer kommt es an: Empirische Evidenzen für Zusammenhänge zwischen Lehrerkompetenzen, Lehrerhandeln und dem Lernen der Schüler [It's the teacher who matters: Empirical evidence of links between teacher competence, teacher performance, and student learning outcomes]. In C. Allemann-Ghionda & E. Terhart (Eds.), *Zeitschrift für Pädagogik Beiheft: Vol. 51. Kompetenzen und Kompetenzentwicklung von Lehrerinnen und Lehrern* (pp. 47–70). Beltz.

Lipowsky, F. (2010). Lernen im Beruf – Empirische Befunde zur Wirksamkeit von Lehrerfortbildung [In-service learning: Empirical findings on the effectiveness of teacher education]. In F. H. Müller, A. Eichenberger, M. Lüders, & J. Mayr (Eds.), *Lehrerinnen und Lehrer lernen: Konzepte und Befunde zur Lehrerfortbildung* (pp. 51–70). Waxmann.

Martin, M., Farrell, M., Seidel, T., Rieß, W., Könings, K. D., van Merriënboer, J. J. G., & Renkl, A. (2022). Focused self-explanation prompts and segmenting foster pre-service teachers' professional vision - but only during training! *International Journal of Educational Technology in Higher Education, 19*(1), 1–29. https://doi.org/10.1186/s41239-022-00331-z

Mason, J. (2002). *Researching your own practice: The discipline of noticing*. Routledge. http://oro.open.ac.uk/790/

Mason, J. (2011). Noticing: Roots and branches. In M. G. Sherin, V. R. Jacobs, & R. A. Philipp (Eds.), *Studies in mathematical thinking and learning. Mathematics teacher noticing* (pp. 35–50). Routledge.

Mayr, J. (2014). Der Persönlichkeitsansatz in der Forschung zum Lehrerberuf [The personality approach to research on the teaching profession]. In E. Terhart, H. Bennewitz, & M. Rothland (Eds.), *Handbuch der Forschung zum Lehrerberuf* (pp. 189–215). Waxmann.

McClelland, D. C. (1973). Testing for competence rather than for "intelligence". *American Psychologist, 28*(1), 1–14.

Meschede, N., Fiebranz, A., Möller, K., & Steffensky, M. (2017). Teachers' professional vision, pedagogical content knowledge and beliefs: On its relation and differences between pre-service and in-service teachers. *Teaching and Teacher Education, 66*, 158–170. https://doi.org/10.1016/j.tate.2017.04.010

Metsäpelto, R.-L., Poikkeus, A.-M., Heikkilä, M., Husu, J [Jukka], Laine, A., Lappalainen, K., Lähteenmäki, M., Mikkilä-Erdmann, M., Warinowski, A., Iiskala, T., Hangelin, S., Harmoinen, S., Holmström, A., Kyrö-Ämmälä, O., Lehesvuori, S., Mankki, V., & Suvilehto, P. (2021). A multidimensional adapted process model of teaching. *Educational Assessment, Evaluation and Accountability*, 1–30. https://doi.org/10.1007/s11092-021-09373-9

Palmer, D. J., Stough, L. M., Burdenski, T. K., JR., & Gonzales, M. (2005). Identifying teacher expertise: An examination of researchers' decision making. *Educational Psychologist, 40*(1), 13–25. https://doi.org/10.1207/S15326985EP4001_2

Piwowar, V., Barth, V. L., Ophardt, D., & Thiel, F. (2018). Evidence-based scripted videos on handling student misbehavior: the development and evaluation of video cases for teacher education. *Professional Development in Education, 44*(3), 369–384. https://doi.org/10.1080/19415257.2017.1316299

Rasch, G. (1960). Probabilistic models for some intelligence and attainment tests. Danmarks pædagogiske Institut.

Richter, E., & Richter, D. (2020). Fort- und Weiterbildung von Lehrpersonen [Professional development of in-service teachers]. In C. Cramer, J. König, M. Rothland, & S. Blömeke (Eds.), *utb Professionsforschung. Handbuch Lehrerinnen- und Lehrerbildung* (pp. 345–353).

Sabers, D. S., Cushing, K. S., & Berliner, D. C. (1991). Differences among teachers in a task characterized by simultaneity, multidimensional, and immediacy. *American Educational Research Journal, 28*(1), 63–88. https://doi.org/10.3102/00028312028001063

Santagata, R., König, J., Scheiner, T., Nguyen, H., Adleff, A.-K., Yang, X., & Kaiser, G. (2021). Mathematics teacher learning to notice: A systematic review of studies of video-supported teacher education. *ZDM Mathematics Education, 53*, 119–134. https://doi.org/10.1007/s11858-020-01216-z

Santagata, R., & Yeh, C. (2016). The role of perception, interpretation, and decision making in the development of beginning teachers' competence. *ZDM Mathematics Education, 48*(1-2), 153–165.

Schack, E. O., Fisher, M. H., & Wilhelm, J. A. (Eds.). (2017). *Research in Mathematics Education. Teacher noticing: Bridging and broadening perspectives, contexts, and frameworks*. Springer International Publishing. https://doi.org/10.1007/978-3-319-46753-5

Scheiner, T. (2021). Towards a more comprehensive model of teacher noticing. *ZDM – Mathematics Education, 53*(1), 85–94. https://doi.org/10.1007/s11858-020-01202-5

Scheiner, T., & Kaiser, G. (2023). Etablished and emerging theoretical perspectives on teacher noticing. In M. Ayalon, B. Koichu, R. Leikin, L. Rubel, & M. Tabach (Eds.), *Proceedings of the 46th conference of the International Group for the Psychology of Mathematics Education* (Vol. 1, pp. 104–108). PME.

Schlesinger, L., Jentsch, A., Kaiser, G., König, J., & Blömeke, S. (2018). Subject-specific characteristics of instructional quality in mathematics education. *ZDM – Mathematics Education, 50*(3), 475–490. https://doi.org/10.1007/s11858-018-0917-5

Seidel, T., & Stürmer, K. (2014). Modeling and measuring the structure of professional vision in preservice teachers. *American Educational Research Journal, 51*(4), 739–771. https://doi.org/10.3102/0002831214531321

Shavelson, R. J. (2010). On the measurement of competency. *Empirical Research in Vocational Education and Training, 2*(1), 41–63.

Sherin, B., & Star, J. R. (2011). Reflections on the study of teacher noticing. In M. G. Sherin, V. R. Jacobs, & R. A. Philipp (Eds.), *Studies in mathematical thinking and learning. Mathematics teacher noticing* (66–78). Routledge.

Sherin, M. G. (2007). The development of teachers' professional vision in video clubs. In R. Goldman, R. Pea, B. Barron, & S. Derry (Eds.), *Video Research in the Learning Sciences* (pp. 383–395). Erlbaum.

Sherin, M. G., Jacobs, V. R., & Philipp, R. A. (Eds.). (2011a). *Studies in mathematical thinking and learning. Mathematics teacher noticing*. Routledge.

Sherin, M. G., Jacobs, V. R., & Philipp, R. A. (2011b). Situating the study of teacher noticing. In M. G. Sherin, V. R. Jacobs, & R. A. Philipp (Eds.), *Studies in mathematical thinking and learning. Mathematics teacher noticing* (pp. 3–13). Routledge.

Sherin, M. G., Russ, R. S., & Colestock, A. A. (2011). Accessing mathematics teachers' in-the-moment noticing. In M. G. Sherin, V. R. Jacobs, & R. A. Philipp (Eds.), *Studies in mathematical thinking and learning. Mathematics teacher noticing* (pp. 79–94). Routledge.

Shulman, L. S. (1986). Those who understand: Knowledge growth in teaching. *Educational Researcher, 15*(2), 4–14.

Shulman, L. S. (1987). Knowledge and teaching: Foundations of the New Reform. *Harvard Educational Review, 57*(1), 1–23. https://doi.org/10.17763/haer.57.1.j463w79r56455411

Shulman, L. S. (1998). Theory, practice, and the education of professionals. *The Elementary School Journal, 98*(5), 511–526.

Stahnke, R., Schueler, S., & Roesken-Winter, B. (2016). Teachers' perception, interpretation, and decision-making: A systematic review of empirical mathematics education research. *ZDM Mathematics Education, 48*(1-2), 1–27. https://doi.org/10.1007/s11858-016-0775-y

Star, J. R., & Strickland, S. K. (2008). Learning to observe: Using video to improve preservice mathematics teachers' ability to notice. *Journal of Mathematics Teacher Education, 11*(2), 107–125. https://doi.org/10.1007/s10857-007-9063-7

Sternberg, R. J., & Horvath, J. A. (1995). A Prototype View of Expert Teaching. *Educational Researcher, 24*(6), 9–17. https://doi.org/10.3102/0013189X024006009

Stigler, J. W., & Miller, K. F. (2018). Expertise and expert performance in teaching. In K. A. Ericsson, R. R. Hoffman, A. Kozbelt, & M. A. Williams (Eds.), *The Cambridge handbook of expertise and expert performance* (pp. 431–452). Cambridge University Press. https://doi.org/10.1017/9781316480748.024

Tatto, M. T. (Ed.). (2013). *The Teacher Education and Development Study in Mathematics (TEDS-M) - Technical report: Policy, practice, and readiness to teach primary and secondary mathematics in 17 countries.* International Association for the Evaluation of Educational Achievement (IEA).

Tatto, M. T., Peck, R., Schwille, J., Bankov, K., Senk, S., Rodriguez, M., Ingvarson, L., Reckase, M., & Rowley, G. (2012). *Policy, practice, and readiness to teach primary and secondary mathematics in 17 countries: Findings from the IEA Teacher Education and Development Study in Mathematics (TEDS-M).* International Association for the Evaluation of Educational Achievement (IEA).

Terhart, E. (2021). Teacher education in Germany: Historical development, status, reforms and challenges. In J. C.-K. Lee & T. Ehmke (Eds.), *Asia-Europe education dialogue. Quality in teacher education and professional development: Chinese and German perspectives* (pp. 44–56). Routledge Taylor & Francis Group.

Terhart, E., Bennewitz, H., & Rothland, M. (Eds.). (2014). *Handbuch der Forschung zum Lehrerberuf* [Handbook of research on the teaching profession]. Waxmann.

Thomas, J., Dueber, D., Fisher, M., Jong, C., & Schack, E. O. (2020). Professional Noticing into Practice: An Examination of Inservice Teachers' Conceptions and Enactment. *Investigations in Mathematics Learning, 12*(2), 110–123. https://doi.org/10.1080/19477503.2019.1681834

United Nations (Ed.). (2015). *THE 17 GOALS | Sustainable Development.* https://sdgs.un.org/goals

United Nations Educational, Scientific and Cultural Organization (Ed.). (2017). *Education for sustainable development goals: Learning Objectives.* https://www.unesco.de/sites/default/files/2018-08/unesco_education_for_sustainable_development_goals.pdf

van Es, E. A. (2011). A framework for learning to notice student thinking. In M. G. Sherin, V. R. Jacobs, & R. A. Philipp (Eds.), *Studies in mathematical thinking and learning. Mathematics teacher noticing* (pp. 134–151). Routledge.

van Es, E. A., & Sherin, M. G. (2002). Learning to notice: Scaffolding new teachers' interpretations of classroom interactions. *Journal of Technology and Teacher Education, 10,* 571–596.

van Es, E. A., & Sherin, M. G. (2008). Mathematics teachers' "learning to notice" in the context of a video club. *Teaching and Teacher Education, 24*(2), 244–276. https://doi.org/10.1016/j.tate.2006.11.005

van Es, E. A., & Sherin, M. G. (2021). Expanding on prior conceptualizations of teacher noticing. *ZDM Mathematics Education*(53), 17–27. https://doi.org/10.1007/s11858-020-01211-4

Voss, T., Kleickmann, T., Kunter, M., & Hachfeld, A. (2011). Überzeugungen von Mathematiklehrkräften [Beliefs of mathematics teachers]. In M. Kunter, J. Baumert, W. Blum, U. Klusmann, S. Krauss, & M. Neubrand (Eds.), *Professionelle Kompetenz von Lehrkräften: Ergebnisse des Forschungsprogramms COACTIV* (pp. 235–257). Waxmann.

Wei, Y., Zhang, Q., Guo, J., & Chen, M. (2023). Learning to teach through noticing: a bibliometric review of teacher noticing research in mathematics education during 2006–2021.

Humanities and Social Sciences Communications, 10(1). https://doi.org/10.1057/s41599-023-01718-7
Weinert, F. E. (2001). Concept of Competence: A conceptual clarification. In D. S. Rychen & L. H. Salganik (Eds.), *Defining and selecting key competencies* (pp. 45–65). Hogrefe & Huber Publishers.
Wermke, W. (2011). Continuing professional development in context: teachers' continuing professional development culture in Germany and Sweden. *Professional Development in Education, 37*(5), 665–683. https://doi.org/10.1080/19415257.2010.533573
Weyers, J., Felske, C., & König, J. (2022). Die Bedeutung von Lerngelegenheiten für das Erreichen von Kompetenzniveaus im pädagogischen Wissen bei angehenden Mathematiklehrkräften [The relevance of opportunities to learn for the achievement of competence in general pedagogical knowledge of pre-service mathematics teachers]. In N. Buchholtz, B. Schwarz, & K. Vorhölter (Eds.), *Initiationen mathematikdidaktischer Forschung* (pp. 247–266). Springer Spektrum.
Weyers, J., König, J., Rott, B., Greefrath, G., Vorhölter, K., & Kaiser, G. (2023). Mathematics teachers' professional noticing: Transfer of a video-based competence assessment instrument into teacher education for evaluation purposes. *Zeitschrift Für Erziehungswissenschaft*, 1–26. https://doi.org/10.1007/s11618-023-01159-7
Weyers, J., König, J., Santagata, R., Scheiner, T., & Kaiser, G. (2023). Measuring teacher noticing: A scoping review of standardized instruments. *Teaching and Teacher Education, 122*. https://doi.org/10.1016/j.tate.2022.103970
Winegard, B [Bo], Winegard, B [Benjamin], & Geary, D. C. (2018). The evolution of expertise. In K. A. Ericsson, R. R. Hoffman, A. Kozbelt, & M. A. Williams (Eds.), *The Cambridge handbook of expertise and expert performance* (pp. 40–48). Cambridge University Press.
Yang, X., & Kaiser, G. (2022). The impact of mathematics teachers' professional competence on instructional quality and students' mathematics learning outcomes. *Current Opinion in Behavioral Sciences, 48*, 101225. https://doi.org/10.1016/j.cobeha.2022.101225
Yang, X., Kaiser, G., König, J., & Blömeke, S. (2020). Relationship between pre-service mathematics teachers' knowledge, beliefs and instructional practices in China. *ZDM – Mathematics Education, 52*(2), 281–294. https://doi.org/10.1007/s11858-020-01145-x
Yang, X., Kaiser, G., König, J., & Blömeke, S. (2021). Relationship between Chinese mathematics teachers' knowledge and their professional noticing. *International Journal of Science and Mathematics Education*, 1–23. https://doi.org/10.1007/s10763-020-10089-3
Yang, X., König, J., & Kaiser, G. (2021). Growth of professional noticing of mathematics teachers: A comparative study of Chinese teachers noticing with different teaching experience. *ZDM Mathematics Education, 53*, 29–42. https://doi.org/10.1007/s11858-020-01217-y
Zhang, Q., Wei, Y., & Liang, J. (2023). Elementary preservice teachers' noticing of exemplary lessons: A comparison of noticing frameworks. In M. Ayalon, B. Koichu, R. Leikin, L. Rubel, & M. Tabach (Eds.), *Proceedings of the 46th conference of the International Group for the Psychology of Mathematics Education*. PME.

The manufacturer's authorised representative in the EU is Springer Nature Customer Service Centre GmbH, Europaplatz 3, 69115 Heidelberg, Germany. If you have any concerns regarding our products, please contact ProductSafety@springernature.com

Printed and bound by CPI Group (UK) Ltd, Croydon, CR0 4YY
02/10/2026
02227140-0001